LINCOLN CHRISTIAN COLLEGE A

PRAISE FOR

BREAKTHROUGH CELL GROUPS

"I have had the privilege of being Karen Hurston's senior pastor for quite a number of years. The Hurstons are synonymous with cell ministry. Karen is truly an expert on the subject of church growth and has twenty-first century insights that bring fresh life to cell ministry in any size church. In her new book *Breakthrough Cell Groups,* she raises the bar for ministries all over the globe to effectively reach their generation."

JOHN KILPATRICK
Senior Pastor of Brownsville Assembly of God, Pensacola, FL

"In this book, Karen shows us how we can break through to reach today's people through creative cell groups. Many years ago, I brought Karen to Phoenix First Assembly to help us develop and start an effective deacons' visitation ministry, whose impact we still enjoy to this day. May the practical insights from this book help you develop an effective cell group ministry that will last until Jesus comes again!"

TOMMY BARNETT
Senior Pastor of Phoenix First Assembly

"Any Christian who yearns to know the dynamics behind the explosive growth of the first century church reads the book of Acts regularly. Any Christian who yearns to know how those dynamics can be repeated for the explosive growth of the twenty-first century church should read this book. Insights gained can be applied in any church, helping the church to grow."

JOHN KIESCHNICK
Senior Pastor of Gloria Dei Lutheran Church, Houston, TX

"For many months I have been saying that Victory Christian Center is the premiere cell church in America. I have made several trips to study the model. It endorses what I have been saying since 1974 — that every cell group MUST have a target! Please do more than read this book: put it into action. Copy their strategy and their spirit and form target groups for every cell — and see the converts flow in!"

RALPH W. NEIGHBOUR, JR.
Houston, TX

"I highly recommend Karen's book, *Breakthrough Cell Groups.* She has found a dynamic cell system that had created groups that reached and broke through to every segment of their city. Read this book and learn how you, too, can create breakthrough cells groups."

DR. DAVID YONGGI CHO

Senior Pastor of Seoul, Yoido Full Gospel Church

"Whether speaking or writing, Karen brings enthusiasm, energy, and excitement that is contagious. Although she could discuss theory with the best of cell church experts, she is fundamentally a practitioner. This book provides the 'tracks to run on' for the person who is serious about the New Testamentmodel of authentic fellowship."

DR. JAMES GARLOW

Senior Pastor of Skyline Wesleyan Church, San Diego, CA

"Karen Hurston gives a detailed account of the history of small group ministry at Victory Christian Center that includes the pitfalls and successes along the way. Her account of what has happened at Victory since 1983, along with a detailed description of their small group ministry is complete with sample group meetings and much more. Very comprehensive and thorough."

TED HAGGARD

Senior Pastor of New Life Church, Colorado Springs, CO

"Karen Hurston is one of the foremost authorities on successful cell group church development. I believe this book will greatly enhance the ability of every pastor to bring his church in line with New Testament methodology of saints doing the work of the ministry, not just the ministers. I highly recommend this book."

BISHOP KEITH A. BUTLER

Pastor of Word of Faith International Christian Center, Southfield, MI

"Karen Hurston's description of Victory Christian Center is exciting and motivating. In careful detail, she describes their extensive network of small groups or cells and the practices that support them. The outcomes are reported in inspiring stories of people who are dramatically affected. She whets your appetite to see God's power released to change lives for the better! Thank you, Karen, for your labor of love in documenting the still unfolding story of Victory Christian Center."

CARL GEORGE

Author and director of Consulting for Growth, Diamond Bar, CA

Breakthrough Cell Groups

BREAKTHROUGH CELL GROUPS

How One American Church Reaches People for Christ through Creative Small Groups

KAREN HURSTON

TOUCH PUBLICATIONS
Houston, Texas, U.S.A.

Published by TOUCH Publications
P.O. Box 19888
Houston, Texas, 77224-9888, U.S.A.
(281) 497-7901 • Fax (281) 497-0904

Copyright © 2001 by Karen Hurston

All rights reserved. No part of this publication may be reproduced, stored in a retrieval system, or transmitted, in any form or by any means, electronic, mechanical, photocopying, recording, or otherwise, without the prior written permission of the publisher. Printed in the United States of America.

Cover design by Don Bleyl
Text design by Rick Chandler
Editing by Scott Boren

International Standard Book Number: 1-880828-31-6

All Scripture quotations, unless otherwise indicated, are from the *King James Version.* Authorized King James Version.

Other versions used are:
AMP— Scripture taken from THE AMPLIFIED BIBLE, the Amplified New Testament copyright © 1958, 1987 by the Lockman Foundation. Used by permission.
NIV— New International Version®, Copyright © 1973, 1978, 1984 by International Bible Society. Used by permission.
NKJV— Scripture taken from the *New King James Version.* Copyright © 1979, 1980, 1982 by Thomas Nelson, Inc. Used by permission. All rights reserved.

TOUCH Publications is the book-publishing division of TOUCH Outreach Ministries, a resource and consulting ministry for churches with a vision for cell-based local church structure.

Find us on the World Wide Web at
http://www.touchusa.org
http://hurstonministries.org

This book is dedicated to my parents,
John and Maxine Hurston.

Dad, *you have been an avid champion of the cell church since 1964 when working with Dr. Cho to start his cells. You are a marvelous mentor, a gifted minister, a wise advisor, and the best father a girl could have.*

Mom, *you are an extraordinary example of a godly wife, mother, and missionary. You are my most cherished friend, my favorite prayer partner, and the wind beneath my sails. Remember: the best is yet ahead! I love you!*

100582

Acknowledgments

Billy Joe and Sharon Daugherty, thank you for the privilege of researching and writing about your great church and exciting cell system. The cell church movement owes you a debt of gratitude!

Jerry and Lynn Popenhagen, thank you for the time and effort you so graciously gave to set up interviews, answer endless e-mails, and tirelessly review and correct manuscripts. I could never have written this book without you (and Tonia)!

Scott Boren, you are a gifted editor and a delight to work with. Thank you and TOUCH for believing in this book enough to publish it.

Victory's dedicated staff, I extend my heartfelt gratitude for time in interviews and observation, including such people as: Bruce Edwards; Steve Worley; Margaret Hawthorne; Terry Glaze; Gary Stanislawski; Tom Dillingham; Jeremy and Carissa Baker; Jodi Hill; C.J. Jacobs; Ed Brownfield; Howard and Cheryl Shouse; Eric and Melody Castrellon; Al and Maria Leerdam; Jerry and Jan Hauser; Henry and Tonia Barlett; Don and Arlene Hanson; Edwin and Delia Miranda; Barbra Smith.

Victory's capable cell leaders and coordinators, thank you for allowing me to observe and ask questions about your groups, including: Pam

Cornwell; Nancy Mashburn; Jim DePriest; Don and Susan Lipke; Allen and Kandi Thurman; Kaye Freeman; Steve and Wendy Pogue; Bill and Michelle Cooley; Leanna Harvey; Les and Tammie Wallace.

The many I met through the conference, phone, audio tape or e-mail, thanks for allowing me to quote you and tell your stories: Rod and Gloria Baker; Charles and Margaret Hodge; Harry and Nikki Latham; Ken Weaver; Bob and Shirley Morton; Tracey Robertson; Nadina Stevenson; Jennifer Lamb; Radhika Mittapalli; Christina Grimm; Wendy Tyler; Jerrylene Birchett.

The many who spent endless hours to tabulate our survey, I am grateful, with deepest thanks to Patricia Towles and Barbara Lassiter.

Don and Susan Lipke, thank you not only for your kind hospitality and time, but also for sharing your group, your family, and your lives.

Doreen Tollis, you are a secretary and office manager of wonderful patience, persistence, and prayerful-ness. I am grateful to God for you!

Dad and Mom, thank you for every chapter you read and critiqued, and for your ongoing encouragement and prayer. Dad, this book was really your idea!

Jesus, most of all, I am grateful to You. You lived and died so every good thing that happened in this book would be possible. Empower us, by Your Spirit, with the boldness and wisdom to break through and reach people for Christ through creative cell groups!

Contents

FOREWORD

Karen lives and breathes cell ministry. She came to Victory to see what was happening with us. Her kindness and sincerity won the hearts of our staff. She asked lots of hard questions and uncovered areas we needed to improve. Her research was thorough in regards to what we are doing. God used this time to strengthen our ministry.

As you read the story of Victory's cell groups, you will sense our struggle. It has been, and continues to be, a process for us. We are pressing ahead with the cell vision. Lives are being changed one at a time. We are grateful to Karen for helping us see that what we are doing can work for others who know they need cells, but just have not found a way to do it.

Billy Joe Daugherty
Senior Pastor
Victory Christian Center
Tulsa, Oklahoma

INTRODUCTION

Join me on a written journey to one of the best-kept secrets among America's cell churches, Victory Christian Center in Tulsa, Oklahoma. "Victory," as its members affectionately call it, has had cells for nearly 20 years. But Victory's leaders are cautious: they wanted their cell groups to be genuinely workable before sharing with others.

What you will discover at Victory is the most evangelistic cell system you have yet seen in America. Last year, Victory's more than 850 cell groups recorded 6,149 salvations; 70% of these salvations came through Victory's Kidz Clubs (children's cells), bus route cells, and S.O.U.L. outreach youth cells. Each month during 2000, Victory's cells averaged a reported 512 salvations. By January of 2001, Victory reported more than 900 cell groups, expecting to soon march past 1,000 cell groups.

You might be like me. My mother tells me I am a lot like her, a woman of great passion. The more I studied Victory, the more excited I became about what I had learned. After months of interaction, I also grew to love Victory's people and leaders. Don't get me wrong; there are several good cell church models, each one worthy of study. But at this point, I haven't gone beyond Victory.

I'm not sure when I fell in love with Victory. Maybe it was during the faith-filled sermons of Pastor Billy Joe Daugherty, hearing his encouragement that everyone get involved in cells. Or while listening to all the inspiring workshop tapes of their first national cell conference, tapes that are the source of a few quotes found later. Maybe it was when I stood before more than 20 volunteers as we manually tabulated cell survey results, with some findings in this book. Or when I went on a "prayer journey" with the Care Pastors to one of their districts, or interviewed Jerry and Lynn Popenhagen, staff pastors and directors of the Pastoral Care Department, or the other staff I talked with.

Maybe it happened when I observed Don and Susan Lipke's zip code cell, or Allen and Kandi Thurman's business cell, or Kaye Freeman's Kidz Club, or Henry and Tonia Bartlett's Friday night international cell, or saw a Sunday Cell for inner city children, or maybe when I talked to the captain of a bus route cell.

Whenever it happened, I fell in love with Victory and remain so, for this reason above all: the staff and congregation love God, His Word, and the lost, and they use the vehicle of cells to make His love known.

Some cell church advocates might argue that Victory Christian Center is not a true cell church because certain cell methods vary from cell group to cell group. But that is the Victory way — different strokes for different folks.

To prepare for our journey, I have spent months observing, interviewing, reviewing questionnaires, and listening to training tapes to discover what makes the Victory cell groups effective. In this book, I include more details than is usual, because I want our journey to be complete. You will go with me into the offices, homes, schools, and lives of the people of Victory. My goal is to show you what goes on, so you can catch a glimpse of those things that stirred my excitement about this church.

My approach is up-close and personal, not analytical. I share lots of testimonies and many names of the people that minister through cell groups. After all, groups are about people, each with a name and a story to tell. I explain how different groups work so you can understand what they do. And finally, I chronicle a day with one of the Care Pastors to reveal an insider's point of view to their ministry.

As you go with me on this journey to Victory, what can you learn? Learn about five breakthroughs to benefit you in your cell ministry.

1. Learn one example of breakthrough cell church success in America. It is great to know about thriving cell churches in Seoul, Singapore, Abidjan, and Bogotá, yet our hearts hunger to hear more of strong cell churches in our own land.

Victory is one such church. As you read the exciting and sometimes "bumpy" journey of Victory's cell system in chapter two, be encouraged. The church had its failures and frustrations, but continued to believe God that cell groups were their destiny. They are now one of the select churches in the world who will soon break through the 1,000 cell group mark.

2. Discover workable breakthrough principles, practices and stories you can share and teach others. Don't go on this reading journey for yourself alone, but also for the many you lead and influence. Observe and use the steps, stories, and testimonies in the sidebars throughout. Notice Victory's approach to leadership training in chapter five. Read the informative insights listed in chapter eight. Glean dozens of "cell church nuggets" to benefit you and those you serve.

3. Learn creative ways the cell church can break through to reach more people. Pastor Daugherty determined that Victory's cell system would be evangelistic, and he challenges each cell to pray for and reach the lost. Through the years, Victory has blended four different group models, all with one goal: to reach even more people for Jesus.

Victory started with home-based geographical groups, inspired by Dr. David Yonggi Cho's church in Seoul, Korea. That later developed into what is termed a "5x5" system, ideally with every five groups overseen by a senior lay leader (area coordinator) and five area coordinators overseen by a pastoral staff member. Although their growth and impact was good, Victory found several pockets of people they could not reach. To touch even more lives, Victory added "target cells" to reach specific groups of people. Soon target cells met weekly in factories, businesses, schools, and workplaces around the city.

Then Daugherty decided that each department of the church should develop its own cells. Victory's leaders discovered many on-site church groups were not finding the same level of relational community, ministry, and care as in their cell groups. They adapted certain aspects of the "meta-church" model espoused by Carl George, and soon every department had cells; even on-site church groups discovered the joy of regular meetings and relationships aimed at outreach, ministry, and care.

Victory most recently began transitioning to G-12, groups of 12 or "principle of 12," a system aggressive in raising up new leaders that started in César Castellanos's church in Bogotá, Colombia. Victory is now busy helping other leadership groups in its system adopt the G-12 approach, a key feature in their goal of 2,000 future cells.

As you read chapters three and four, learn about 24 of the 35 types of groups Victory uses to impact a broad range of people. Consider how you, too, can help your group and cell church break through and reach even more for our Lord!

4. Discover a new way to break through in ministry to children in the cell church. Parents are concerned for their children. Rarely have I taught in a leaders' conference where no one asked about the role of children. As you read about Victory's fast-paced and inspiring "Kidz Clubs" in chapter four, discover a dynamic new way to minister to children, from the department that has the most off-site children's cells of any church in the nation.

5. *Use this book as curriculum for a group or committee to experience breakthrough in their thinking about the cell church.* At the end of each chapter are small group discussion questions, complete with an "icebreaker." If you lead a group or committee that is looking into cell groups or the cell church, material in this book might be your answer to prayer for your own breakthrough or turning point.

Curious? Intrigued? Read on and I think you might fall in love as well . . . or at least find your concept of cells broadened and your heart greatly warmed.

But a word of caution. One beauty of the cell church is that it is constantly changing in order to meet the needs of those it seeks to serve and reach more effectively. No matter what you learn or glean from this book, know that its statistics are "time dated," and will expire. As of January of 2001, Victory has 913 cells with 35 varieties of groups. Even as I was writing, Lynn Popenhagen told me that Victory will soon be reaching 1,000 groups, and that one of their beloved Care Pastor couples, Howard and Cheryl Shouse, were going to move out of state to work in a family member's church. If you go to Victory, you might find their numbers have increased, and some staff members' names have changed. But one thing will remain the same: Victory will continue to use the vehicles of groups to reach the lost, and to train and minister to believers.

So, read on, and be open to the creative possibilities in using cell groups to break through and reach every segment of your community!

1

Breakthrough Cell Groups

Victory's Cell Challenge

VICTORY'S EVANGELISM GROUP PROFILE FOR YEAR 2000

Total salvations through 850 cells: 6,149

Monthly range of salvations through cells: 451-612

Monthly average of cell salvations: 512

Average annual number of salvations through each cell: 7

Average annual number of salvations through each Kidz Club, bus route cell, and S.O.U.L. Outreach Youth Cell: 23

Average annual number of salvations through other kinds of cells: 3

On a cool Wednesday night in Tulsa, I join 2,000 cell leaders, group members, and church attendees packed into the multi-purpose auditorium on Victory Christian Center's west campus. Sharon Daugherty has just led the ethnically diverse crowd in lively worship, and Senior Pastor Billy Joe Daugherty steps to the platform's center.

Daugherty looks at those gathered, and begins, "Victory has 900 cell groups that meet across the city of Tulsa — in homes, businesses, schools, factories, service and construction sites, the Cancer Treatment Center, downtown high rise office buildings, apartment complexes, and our own church property. If you aren't already, it's time for you to be in a cell group!"

OVERALL PROFILE OF VICTORY CHRISTIAN CENTER (January 2001)

Year started: 1983

Founding pastor: Billy Joe Daugherty

Number of cell groups: 930

Current membership: 11,000+

Sunday worship attendance: 7,700 (9, 10, 11:00 A.M., 6:00 P.M.)

Midweek attendance: 2,200

Average of number weekly attending cells in 2000: 7,483

Highest event attendance: 21,000 (Easter, 2000)

Number on pastoral staff: 31

Number on support staff: 98

Denomination or movement: Charismatic Word Church

Affiliation: Pentecostal and Charismatic Churches of North America (P.C.C.N.A.) & International Charismatic Bible Ministries

Best known identifier: Two of three main campuses across the street from Oral Roberts University

Source: Bruce Edwards, Associate Pastor

Many in the mid-week congregation nod their heads as Daugherty continues his challenge, "God has called us to go into every man's world! If you can't find a cell group you like, maybe it's time for you to start one. Think about having a cell group in the neighborhood where you live, the place where you work or go to school, maybe meeting during lunch time."

Daugherty launches into his sermon, explaining that Jesus Himself had a small group. In Luke 6:12-13, Jesus prayed all night before choosing His group, the twelve disciples. Then, in Acts 2:42-46, large groups of New Testament Christians met both in the temple and in small groups from house to house.

"That pattern continues at our church," Daugherty stresses. "Small groups, or 'cell groups' as we call them at Victory, are vital. A cell is a place you can know someone personally, not just as a face in the crowd. A cell is a group of people where you can be known and cared for, and have practical needs met. In a cell you can ask questions about God's Word and the Christian life, and share out of your heart the deposit of God's Word already in you."

Living Examples

Daugherty's repeated "cell challenge" to his congregation has had great impact. From 1983 to 2001, Victory has grown from 29 groups to breaking the one thousand cell mark, with one of the longest continuous "group track records" of any church in America. By July of 2000, more people (7,344) attended Victory's cell groups than the estimated 7,000 who then attended Victory's four Sunday worship services. While many church members attend cells to aid them in their Christian walk, many others who attend cells are not yet involved in Victory; these cells form Victory's most effective channel for discipleship and evangelism. Even Ralph Neighbour, Jr., a leading communicator in the cell church movement, has publicly declared Victory's cell group system the most mature system he has yet seen in America.

What impact have these groups had? My first research trip to Victory, I heard Daugherty's inspiring challenge to his congregation. On my second research visit, I was overwhelmed by verbal and written testimonies from living examples of Victory's breakthrough cell groups, groups that have broken through to salvation and answered prayer.

Kerry and Angela Wilburn's workplace cell group knows that God answers prayer. Already, members have received jobs and promotions, broken

VICTORY'S CONTEXT: TULSA

Tulsa proper population:
390,437 (July 2000)

Tulsa county population: 551,141

Size of Tulsa proper:
192.3 sq. miles

Size of Tulsa county:
555 sq. miles

Number of churches in Tulsa:
500 *

Typical household:
2.4 persons

Average household income:
$47,444

Unemployment: 2.5%

Source: Metropolitan Tulsa Chamber of Commerce (July 2000)

**Source:* *Tulsa Metropolitan Ministry*

relationships have been restored, and one small child has been healed of cancer.

In Randy Kreil's "zip code cell," David confessed that he was a cocaine addict with family problems. The group rejoiced with David as he tearfully received Jesus as his Savior, then watched his water baptism during Victory's next Wednesday evening service.

Samantha Franklin prayerfully started her workplace group in a tense government department. Fifteen people attended the first meeting, and one lady accepted Jesus right there in her office.

When Robin arrived at Victory in a wheelchair, she and her caretaker Alice were welcomed by Jerry and Shirley McCoy, leaders of a Sunday cell. "I just can't tell you how many churches we have tried and it never worked out," Alice said, with tears in her eyes. "I thank God for our Sunday cell. Robin loves to come, and so do I."

Mikael Axelsson and his wife had just moved from Sweden to Tulsa. They were lonely for friends until they discovered Todd and Charly Young's nearby zip code cell group. Axelsson declares, "Now we want to become cell leaders."

A Place to Belong and Become

Victory's cell system has created a network that gives members a place where they can belong and become who God has called them to be. Consider Wendy Tyler's story. Wendy had never gone to a large church, but her husband insisted they attend Victory. The first Wednesday night they went, they both knew Victory was the church for them and became members. Even though they joined, they still questioned how they would ever find meaningful relationships in such a large church.

The second week after they joined Victory, the Tylers attended a cell group. Two weeks later, they started training to be cell group apprentices. Within six months, the Tyler family led four cells: Wendy and her husband led a zip code cell group, their 10-year-old daughter was leading a Kidz Club, their 17-year-old foster son led a high school cell group, and their 16-year-old foster daughter had a teenagers' cell group in their home.

Radhika Mittapalli, a convert from Hinduism, also wasn't sure where she fit in Victory, and had considered leaving. Then she heard someone talk about cells, and instead started a young professionals' group. Mittapalli no longer questions whether or not she belongs in Victory. She now serves as a volunteer "young adults cell coordinator," and leads a G-12 group that has birthed five other groups, including two workplace cells. As a result of her diligence, Mittapalli has become a key leader among Victory's young single professionals.

A Cell Group Defined

What is the "Victory definition" of a cell? With few exceptions, a Victory cell group involves three or more people who meet together on a weekly basis. Whether it be a "zip code cell" that meets in a home, a workplace cell that meets in an office, a G-12 cell that focuses on leadership, a "Sunday cell" that meets in a church facility, or any of Victory's 35 varieties of cells, each cell has four purposes: 1) to evangelize and reach the lost, for this is the most important priority for any group; 2) to minister and give pastoral care to the needs of its members, both during and between meetings; 3) to disciple members in spiritual growth; and 4) to multiply when it reaches twelve people (or a number past the group's ability to provide care), in order to start other cells.

Victory's cell leaders guide their meetings by Daugherty's "five-fold vision." Groups that can take 90 minutes for a meeting, such as the home-based "zip code cells" and the G-12 groups, do all five parts of the vision. Groups more limited in their time, such as cells that meet during

TYPICAL GROUP PROFILE

Current Attendance:
8 people

One Year Ago: 6 people

Four Months Ago: 6 people

Composition: A wide mixture or the same gender

Meeting Place*: Off site — in a home, apartment, or workplace

TYPICAL GROUP MEETING

Duration: 75-90 minutes

Curriculum: Given by Victory, the weekly lesson is printed in Sunday's bulletin

Refreshments: None

Praise and worship: 10-15 minutes

Teaching: 25-30 minutes

Discussion: 15-20 minutes

Prayer and ministry to one another:
10-15 minutes

Fellowship: Remainder

Source: Representative sample (n=105) survey taken by Karen Hurston, July 2000

**Source: Pastoral Care Department*

a lunch hour in an office setting, must do at least three parts of the vision — typically including prayer, the Word, and fellowship with interaction. This crucial five-fold vision includes:

1. Worship and Praise: This focus on magnifying the Lord and exalting Jesus through praise and worship, often takes ten to fifteen minutes. When a group meets in an office or a setting where song is not possible, worship can include reading a Psalm or Bible passage on God's goodness, or brief words of thanks to God. "Worship," Daugherty teaches, "is much more than a 'preamble' to what happens next. Worship is what we were created to do."

2. Prayer: While a group will take another ten to fifteen minutes to pray for the needs and requests of group participants, Victory also stresses praying for others outside the cell — including the government, pastor and church staff, schools, missionaries, and the body of Christ overall. "Prayer moves the hand of God," Daugherty teaches, "and releases His blessings. Prayer changes the world, changes circumstances, changes us. Prayer also stops the power of the enemy."

3. Word of God: During the next ten to fifteen minutes, the leader is to teach the lesson outline printed in that Sunday's bulletin, based on a former sermon Daugherty preached. Each leader is to add his own illustrations to tailor the truth given for that particular group. Victory repeatedly stresses that this is to be a short lesson on a nugget of truth, not an extensive, in-depth teaching.

4. Fellowship and Interaction: Triggered by discussion questions written at the end of each lesson, these fifteen to twenty minutes give each person in the group an opportunity to share what is on his or her heart, whether that be a question, comment, insight, or testimony. It is the leader's goal to have equal participation during this time.

5. Ministry and Evangelism: Victory encourages leaders to allow the gifts of the Spirit to operate during these closing fifteen to twenty minutes. The leader and his apprentice are to minister, edify, and strengthen those gathered, and to help the group learn to flow in the Spirit's gifts. If unbelievers are gathered, this time can instead be used to answer questions and share God's love. If no unbelievers are present, this time can also be used to plan better ways to reach the lost.

Sold on Cells

Victory's four purposes and five-fold vision have resulted in strong cell groups and a seemingly endless list of answered prayer. How can Victory's cells be so effective? "The senior pastor of a church," Billy Joe Daugherty declares, "must be sold on cells. The reason most churches don't grow is because of the attitude of leadership. It starts between the ears of the pastor, in his mentality and thinking."

Daugherty sees his job as equipping and releasing people, so that they can do the work of the ministry. He continually teaches his congregation that they are to grow from 'babyhood' into maturity. Maturity is proven when a person becomes a reproducer. As Hebrews 5:12 states, one is to be a teacher of others.

Daugherty concludes: "Needs in our congregation will only be met as people become teachers and reproducers of others in cells. Only then will each person fulfill God's plan. Only then can this end time harvest be preserved."

CHAPTER ONE: SMALL GROUP DISCUSSION

Icebreaker: Describe your ideal cell group (size, composition, focus, activities).

Questions about this chapter:

1. Name two key Bible passages and one benefit of group participation listed in this chapter.
2. Listed were several brief cell testimonies or stories. Which one impressed you the most? Why?
3. How do cell groups provide a place "to belong and become"?
4. In your own words, what is Victory's five-fold vision? How does this apply to their cell groups?
5. How important is it for the senior pastor to be "sold" on cells?

Application: Share one principle or practice you gleaned from this chapter. How could that be best integrated into your present or future cell group? Your church?

2

THE JOURNEY

How Victory's Group System Started and Grew

By 1976 the 24-year-old Billy Joe Daugherty already had his list of achievements. He had majored in Christian education at Oral Roberts University, and studied in Rhema Bible Training Center; he and his wife Sharon had served as Christian education directors, as youth pastors, and as traveling ministers. Daugherty thought he would be working with youth, in Christian education, or in a traveling ministry for the rest of his life.

In 1979, Tulsa's 300-member Sheridan Christian Center asked Daugherty to become their senior pastor. Daugherty knew God was in that request, and was glad to pastor the same church he had formerly served as youth pastor. But he soon found himself perplexed. Kenneth Copeland, a well-known Bible teacher, had earlier given a prophecy that Sheridan Christian Center would "explode out of these walls and touch the whole world."

The new senior pastor struggled with the content of Copeland's earlier prophecy. He was comfortable with his church's size, and didn't think growth was needed. After all, Daugherty had been raised in small churches where the pastor did most of the work and ministry, and thought that one should want quality, not quantity. While praying

Daugherty objected, "God, I'm not interested in numbers."

Daugherty was surprised to hear God respond, "I am. Whether you call them numbers, masses, or multitudes, they are all people for whom My Son Jesus died. To say you are not interested in numbers is to say to Tom, Joe, and Sally, 'Go to hell.'"

From the moment he understood God's viewpoint, Daugherty's thinking changed. He finally realized that each of "those numbers" was someone Jesus loved, someone for whom Jesus died. God considered each of "those numbers" precious.

After Daugherty's renewed understanding, growth at Sheridan Christian Center came quickly. Even that first year, the congregation doubled, and then doubled again. To keep up with the growth, Daugherty struggled to minister personally to all those coming, working fourteen hours a day, seven days a week.

Then it happened. Just walking 50 yards across a junior high athletic field, Daugherty could not catch his breath. At that moment he knew something was wrong. The stress and the pressure were too much for him. "If you keep up this pace in this way," came the thought, "you're not going to live very long."

By 1980 the 28-year-old overworked Billy Joe Daugherty was desperate. How could he possibly care for the growing numbers God kept bringing to the church?

The First Home Groups

Daugherty knew he had to have help and share pastoral care with others. He had heard about other churches starting home groups and thought it was a good idea. So the next Sunday at church Daugherty declared, "We're going to start home groups. How many of you would like help?"

A few hands went up. Daugherty then invited those 12-15 people to meet with him that evening at one side of the auditorium. At that brief meeting he gave those first group leaders the only "training" they received: "I want you to meet with people once a week in your homes and teach them the Bible. Pray with the people and love them. Are there any questions?"

There were three or four questions, then Daugherty prayed. After a total of 20 minutes of "preparation," he released his first crop of leaders to start groups.

Problems soon arose. Sheridan was an interdenominational church with people from different backgrounds, each with different opinions and doctrines. "During those next two years," Daugherty shook his head as he reflected, "some of the worst things one could imagine happened in those home groups. Some groups cast devils out of people and others put devils into people. Some taught 'revelations' no one had heard before or since."

First Groups Disbanded

After a year of frustration, Daugherty disbanded those first groups. For the next two years he did not even consider starting groups again in his mainly white congregation.

But the needs of his congregation continued to grow. Because of several problems with expansion on the same site, Daugherty decided to start a church at a new site, leaving behind those who wanted to stay. In 1981 Victory Christian Center began with a core of 1,600 people.

Even though Daugherty's first try at groups had not gone well, he knew he must still do something to provide better pastoral care. Like Moses in Exodus 18:13-26 and the disciples in Acts 6:1-7, Daugherty knew that he must delegate ministry to lay leaders. Only in that way could the growing congregation's needs be met. So Daugherty began to pray afresh and look for material on groups.

THE TURNING POINT

In 1983, Daugherty attended a church growth conference held at Dr. David Yonggi Cho's vast and growing church in Seoul, Korea. While hearing about and observing the extensive group system there, he knew that groups were an important part of God's destiny for Victory. Daugherty also began to understand several key elements: he needed a clear cut plan for a group system; he needed good training for potential leaders; and he needed one person or couple to oversee the groups and hold them accountable. Daugherty also saw the importance of providing a weekly lesson that issued from his heart and thinking. What was shared in the groups needed to be in harmony with the larger church vision.

After he returned from Korea, Daugherty began a process of training and preparing to restart groups. He spotted Jerry and Lynn Popenhagen, a sharp couple, who had been discipled in a group and understood how groups worked, and asked them to come on staff to head up the groups.

FAMILY MINISTRY LUNCHEONS

Victory started "Family Ministry Luncheons," potlucks (termed "potfaiths" or "potblessings") held the third Sunday each month in more than 100 homes around the city. After the third Sunday morning church service, different staff members would go to various homes. During these pivotal luncheons, the staff talked to members about the importance of small groups that met in homes. Even today many Victory staff members recall darting to a store to buy food to take, anxious to meet and share the benefits of home groups with their precious members.

When Jerry Popenhagen joined the pastoral staff to head the

groups, he resigned from a lucrative job at American Airlines, taking a 50% cut in pay. During that time, Daugherty and the Popenhagens gathered as much information as they could on cells and small groups, with Daugherty deciding the basic direction he wanted the new groups to head. Under Daugherty's direction, the Popenhagens compiled a training manual that included such items as: how the groups were to operate; how to do follow-up; how to track people. To help further change the thinking of the congregation, they purchased and distributed a large quantity of Dr. Cho's *Successful Home Cell Groups*, encouraging anyone interested to read the book.

Bible Fellowships

Daugherty named that second crop of home groups "Bible Fellowships." He gave several months of careful training to 29 leaders. Before, many had thought their only role was to come to church services and help in an area of service. Daugherty highlighted the truth of Ephesians 4:11-12, that those full-time in ministry were to "equip the saints for the work of the ministry."

By the time Daugherty commissioned those 29 leaders to start groups, they understood their roles, knew what to do, and knew what to teach. Group meetings were limited to an hour and a half, and hosts were to have little or no food. Daugherty taught leaders to encourage all members to participate in a group meeting in one way or another: one might greet, another say the opening prayer, still another might lead the worship, another teach, and another take the prayer requests. During group meetings there were to be five basic ministry activities, following Victory's "Five-Fold Vision:" each group was to have worship, prayer, the Word of God, fellowship, and outreach. Groups were challenged first to grow, then to multiply, and finally to divide into two or more groups.

When those groups began, Daugherty had each week's cell lesson printed in the Sunday bulletins. This allowed everyone to see that groups were significant, one of Daugherty's ways of "seeding" vision for groups into his congregation. Victory began to publish a complete monthly listing of every group, with name of leader, meeting day, time, and place. Because those first group listings were limited, they were also printed in the bulletin; later, "cell group directories" became so large they had to be published separately.

Bob and Shirley Morton, involved in the previous monthly Family Ministry Luncheons, went through that training and held the first official "Bible Fellowship" meeting in their home. The Mortons did not consider themselves likely group leaders; they didn't even pray regularly for the lost or for others' needs. Yet, after Daugherty prayed to commission them as leaders, they found unbelievers and sick people coming to their Bible Fellowship — including a satanic high priest with bleeding ulcers. As the Mortons prayed with them, many of those unbelievers came to the Lord, and many sick were healed, including the recently converted satanic high priest. Morton is now Victory's staff pastor over hospital visitation and 'healing cells', and declares, "I know that the results we had in our Bible Fellowship were a direct impartation of our Pastor's prayer for us."

There were some "bumps" in the journey. A month after the Bible Fellowships started, a woman phoned Lynn Popenhagen to complain that the group she attended was a bit strange. "The leader has this big stopwatch," the lady began her explanation. "Our leader figures out how many of us are there, and how much time she's been told a group should spend in fellowship and interaction. As soon as any one of us has talked to the end of his 'allotted' time, she interrupts what he is saying and informs us it's time for the next person to share."

So Daugherty and the Popenhagens stressed to the leaders that they had given them guidelines, not mandates. It became clear that leaders needed both ongoing training and continual encouragement.

Even with a "bump" or two, it wasn't long before that second crop of leaders started seeing results. Lee and Sue Boss had met and married at Victory, and started a Friday night cell group. Even though Lee was a very busy man, a broker who sold real estate and businesses, he and Sue had a vision to grow their group and start others. Over the next three years, the Boss's one group birthed 75 others. Not long after, they felt a call into full-time ministry, and now pastor Victory Christian Center in Austin, Texas. Interestingly enough, most of that second crop of 29 leaders are now in full-time ministry.

Don't Go Fishin' in Your Own Driveway

By 1987 Billy Joe Daugherty was bothered. Victory, located in one of Tulsa's rich suburban areas, had grown well. Victory had climbed to 204 groups in 1985, and to 450 groups in 1986. But even with all of the good things that had happened, something was wrong. Daugherty was bothered by his passive but well-managed staff and by Victory's content, well-dressed, and mostly white congregation.

Daugherty called his staff together, and began his challenge: "I see in Luke 10:2 and John 4:35, that there are fields around us that are ripe and ready to harvest. You can get a fishing pole," Daugherty explained, "put bait on it, and throw it out on your driveway. It can cast well, and reel in easily, but there won't be any fish on the line. If you want to catch fish, you don't go fishin' in your own driveway. That's what I feel we have been doing."

The staff agreed that the people they were trying to reach were not responding as they once had. Daugherty continued his challenge: "Then where are they? Who are the ripe harvest fields in Tulsa? How can we reach them?"

As the brainstorm began, so did the surprises. The consensus? The most responsive people in Tulsa were in the government projects, especially the children in poor areas.

Some objected. Poor children did not bring large tithes and offerings. They made noise, came in dirty, and required finances just to transport them to church. Besides, their parents would start coming; the cultural and racial mix would change, and some members might leave.

But Daugherty was determined. He began preaching on Jesus' parable of the supper feast in Luke 14:16-24. When the master's invited guests refused to attend his prepared feast, he sent his servant to bring in the poor, the maimed, and the blind. When they did not fill the banquet room, he commanded his servant to go into the "highways and hedges." Daugherty did not just preach on the subject; he and Sharon themselves went into the projects one weekend a month, and preached both Friday night and Saturday afternoon, with a free medical clinic on Saturday morning. Soon Victory sent buses into the projects to pick up children and their parents every Sunday.

Some did leave, but Victory had become fully interracial. Daugherty declared: "Our doors are now open to everyone. We might have people come and sit by you who don't look like you, act like you, talk like you, dress like you, or smell like you. Welcome them anyway. We want to welcome everyone not just to come to Victory, but also to stay."

Target Cells

Victory's cell system expanded with this new and necessary emphasis on diversity. Target cells of all types, which Victory designed to reach a particular type person, quickly became a growing focus. Target cells sprang up in homeless shelters, in American Airlines (to reach mechanics and agents), in downtown high rise buildings, in factories, in athletic involvements, and in different specific professions. Soon "bus route cells" sprang up, as did cells targeting young children and youth in Tulsa's inner city.

Target cell leaders went into new areas as diverse as businesses and factories, with the goal of reaching the people in that area. All types of people attended the newly established "target cells," rich and poor, with some groups reporting participants as different as Catholics, Methodists, Presbyterians, Buddhists, Muslims, and atheists — each given an opportunity to hear the Gospel.

From Growth to Downslide

By the end of 1987, Victory had reached a peak of 508 cell groups. Daugherty's new emphasis on diversity had not only resulted in greater group growth, but had succeeded in reaching people in Tulsa no other church had been able to touch.

It was clear that the Popenhagens could not handle pastoral oversight of the groups alone. Victory's group system required a full department, with five full-time area coordinators and a secretary. Each area coordinator not only ministered to current groups under his charge, but also recruited and trained more leaders. Victory's cells were going well.

But the year 1988 was not a good one for Victory. The fall of two prominent televangelists dealt a devastating blow to many churches, including Victory's congregation and their outreach efforts. They had started a building program, which drained much of Daugherty's attention and focus away from the groups. He rarely mentioned the groups when he spoke. Victory's finances grew tight, and some staff members had to be terminated. The five area coordinators and the secretary were released; the Popenhagens were once again the only ones overseeing Victory's groups. Other departments had lay offs as well, including the two full-time women in the data processing department who had entered all the information into the computerized group tracking system. By the end of that year, the number of cells had dropped to only 275.

In 1990, even though the groups had slid to less than 200, Daugherty decided to hire Care Pastors. Under the Popenhagen's direction, each Care Pastor was to oversee 300 family units and the groups in a district, covering a specific geographical area. "There was no way," Jerry Popenhagen emphasized, "that just the two of us could do all the weddings, funerals, and hospital visits to all those people. We had to have help."

The Popenhagens then oversaw what was termed the "Pastoral Care Department," manned by Care Pastors. It was then that the term "Bible Fellowships" was dropped, and for a short time Victory used the name "Care Groups."

Daugherty's attention, though, was still on the building program. He no longer held groups in priority. By the end of 1990, the entire system slid to 135 groups.

A Heart for the Nations

In 1991, Daugherty felt a burden for Russia, and by August of that year he joined Lester Sumrall, a well-known Bible teacher and pastor of a church in South Bend, Indiana, on a crusade in St. Petersburg. Daugherty already had his booklet "This New Life" translated into Russian and used as a tool to teach new converts about salvation, the Holy Spirit, faith, righteousness, healing, and the abundant life. More than 25,000 Russians gave their lives to Jesus in the seven services of that first crusade, and Daugherty knew God was calling him to return.

Over the next 18 months Daugherty and his wife Sharon went to Russia 16 times, each time holding a crusade and even planting a church and Bible institute in St. Petersburg. Each crusade resulted in thousands responding to the Gospel; in one seven-service crusade in January of 1992, more than 96,000 accepted Jesus as their Savior in altar calls. Within two years, through the crusades, church, and Bible institute, Victory had distributed 3.3 million pieces of Gospel literature, including one million New Testaments, 300,000 complete Bibles, 1.6 million "This New Life" booklets, and more than half a million illustrated children's books.

During that time, God birthed a heart for the nations in Daugherty. Even after this focus on Russia, Daugherty and Sharon still traveled each month to different countries to minister.

How did Daugherty's heart for the nations impact Victory's groups? Many cells now "adopt" one of Victory's 150 missionaries, regularly pray for that missionary and his family, exchange e-mail notes and letters, and even have that missionary speak at the cell when he and his family return to the States.

But there was a negative result. With Daugherty's continuing focus off the groups, their numbers slid even more, down to 96 groups by the end of 1993. At that time, Daugherty again saw the importance of groups, and he began to emphasize groups as Victory's priority once more. He talked about groups when he preached, and he actively encouraged more people to go into group leadership. With Daugherty's correction of priorities, the groups started growing. Even now Daugherty will say, "The key to a healthy cell group system is the emphasis the senior pastor puts on it."

Every Department Develops Cells

By 1995, Daugherty's renewed focus on groups had yielded great increase, and Victory had grown to 403 groups. Care Pastors had expanded their group structure, and placed volunteer area

VICTORY'S FACILITIES USED FOR WORSHIP AND ON SITE GROUPS

THE EAST CAMPUS covers 12 acres with a large 112,000 sq. ft. building, which houses most church offices for staff members overseeing groups, including the Pastoral Care Department. Activities in this building include a wide variety of cell related meetings, such as: cell leaders' training; youth and singles' activities; Sunday inner city children's cells (through the bus ministry); and benevolence, food, and clothing outreach.

THE WEST CAMPUS is 3/4th mile away, across from the sculptured praying hands on the ORU campus. This 160,000 sq. ft. building on 50 acres not only houses classrooms for Victory Christian School's 1500 students and the Sunday 10:00 A.M. and Wednesday 7:00 P.M. worship services, but also classrooms for Sunday cells, Wednesday on site children's cells, music cells, the Saturday "Couples Connection," various workshops, and the executive pastoral offices.

THE MABEE CENTER, located on the ORU campus, is a rising amphitheater-style sports arena that seats 2,500, which can be expanded to seat more than 10,000, and is the rented site of Victory's 9:00 A.M., 11:00 A.M., and 6:00 P.M. Sunday worship services. It is also the site of the church families' children's cell services on Sunday, with a prominent booth filled with cell information.

VICTORY BIBLE INSTITUTE (VBI) is 4 miles west of the west campus, with 110,000 sq. ft. on 5 acres for VBI classrooms and Victory World Missions Training Center. It also houses Victory by Virtue and Pastor Sharon's Bible Study "study cells" on Tuesday and Thursday; inner city youth "S.O.U.L." service on Sunday; and Hispanic service on Sunday afternoon. *All first year VBI students are put in cells led by second year students. These cells are not counted in Victory's monthly listing.*

Source: Steve Worley, Victory's facility director since 1990

coordinators over every three to five groups. Victory's groups were back on the upswing.

Then two related concerns surfaced. First, only the Pastoral Care Department had groups listed in the monthly "group directory." Groups were something that just one department did, not the life of the total church. Secondly, some of the more traditional facility-based groups, like Sunday School classes and age-graded groups, were not having the same level of attention, relationship, ministry, and outreach as the home-based groups. Influenced in part by Carl George's book *Preparing Your Church for the Future,* Victory changed the name of the "Care Groups" to "cell groups," and began to turn even on-site facility-based groups into cells.

Daugherty made the decision that every department would develop its own cells, a big paradigm shift for some department heads to embrace. But soon even Sunday School teachers were trained to be cell group leaders, and began to concentrate on interactive discussion, mutual ministry, and reaching out to others. On-site facility-based groups were then listed in Victory's monthly "cell group directory," and cells became the core channel for ministry outside of Victory's worship services.

Now every department had its own variety of cells. The Pastoral Care Department continued to oversee the home-based zip code cells, business and workplace cells, and others; the Christian Education and Outreach Department had Kidz Clubs, S.O.U.L. Outreach Youth Cells, and bus route cells; the Personal and Family Life Ministry Department oversaw marriage cells and support group cells; six other departments or ministries each oversaw one or two other varieties of cells. By 1997, Victory had 502 cell groups.

G-12

Not long after the addition of departmental cells, Daugherty heard about the "groups of twelve (G-12)" model that sprang out of César Castellanos's International Charismatic Mission. This church in Bogotá, Colombia has 20,000 cell groups and 45,000 in Sunday attendance, and is known as the cell church most aggressive in developing leaders.

Daugherty felt that Victory needed to raise up many more leaders, so in 1998, he started his own G-12 group with key staff leaders. Every week Daugherty teaches the cell lesson in his G-12 cell, followed by discussion, prayer, and ministry; only then does he deal with any "business" or staff tasks. Each of those 12 department leaders lead a group of 12 while currently receiving support, mentoring, and encouragement as a member of Daugherty's leadership G-12.

According to Daugherty, this transition to leadership G-12 groups "has revolutionized and strengthened us. We're still learning and growing in this, and have found it hardest to transition with our staff pastors and department leaders. But even with the difficulties of transition, having G-12 groups has helped create new momentum in our group system."

Bruce Edwards, Victory's associate pastor since 1994, expressed the same dilemma: "This transition was most difficult with staff. The staff thought that turning what they did into G-12 groups was just another thing to do, an additional burden. But Pastor Billy Joe intently showed them that this was meant to help them; in this way, each staff pastor could multiply himself."

Daugherty met with Victory's department heads every week for nine months, sharing his vision for G-12, and before long, even the most reluctant staff members began to understand. Staff members now work their G-12 groups into their regular weekly schedules, many meeting with their leaders in a G-12 during lunchtimes, or the hour and a half before or after lunch. Now each area — whether it be children, youth, singles, ladies, men, districts — has its own leadership G-12, with each staff pastor meeting with his leaders on a weekly or biweekly basis. Unlike the church in Bogotá, Victory has chosen to count its G-12 groups among its cell groups.

RETENTION OF VISITORS

With all the salvations Victory records through their cell groups, the question arises: What about their direct impact on Victory's growth? Because the cells meet throughout Tulsa's sprawling 30-city metro area, it would be unreasonable to expect many of these new converts to attend a distant church. Instead, Victory keeps track of the retention rate of visitors to worship services. According to Terry Glaze, New Members' Pastor, both the number and retention of visitors has greatly increased over the past three years. In 2000 alone, there were 7,190 visitors to Victory, including 3,524 visitors from the metro area, who were unchurched or churched with requests for more information. Of that number, 1,592 (or 45%) became church members. From July of 2000 to January of 2001, Victory increased in worship attendance by an annual rate of 20%. Cells are clearly bringing growth to Victory.

During this time, Victory's departments increased their current cells and developed new varieties of cells. By the end of 1998, Victory had 724 cell groups, climbing to 880 groups in 1999. Now an estimated half of Victory's groups, such as zip code cells, focus on attracting current church members, and helping them in their Christian walk. The other half, such as business cells and Kidz Clubs, are more openly evangelistic, and often involve people who have never even heard of Victory.

Adaptations

Lynn Popenhagen remembers visiting pastors who complained that they had tried cells, but that it "just didn't work for us." Popenhagen responds by saying that there were times cells didn't seem to work at Victory either. But they kept believing cell groups were God's design for Victory, and persisted through the difficult times.

Victory's "blended" group system continues to make central use of its five-fold vision: 1) praise and worship; 2) prayer; 3) the Word of God, woven throughout the cell lesson; 4) fellowship and interaction; 5) ministry and evangelism. Some groups that meet in a restaurant, factory, or place of business might only have twenty to thirty-five minutes to meet, but must do at least three parts of the five-fold vision. Instead of taking fifteen minutes for song, the leader might simply lead the group in statements of worship and adoration to God, and adjust the remainder of the meeting accordingly.

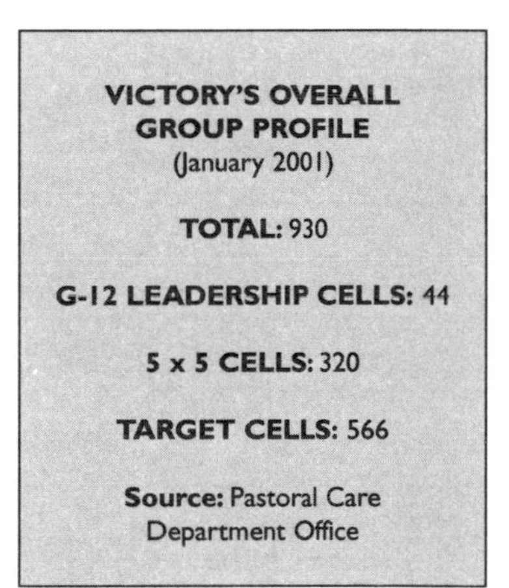
VICTORY'S OVERALL GROUP PROFILE
(January 2001)

TOTAL: 930

G-12 LEADERSHIP CELLS: 44

5 x 5 CELLS: 320

TARGET CELLS: 566

Source: Pastoral Care Department Office

No matter how long or short the meeting, the entire time is to be bathed in prayer. Daugherty's written vision for cell group leaders contains this insight: "The key to success in your cell group ministry will be you, the cell group leader, and prayer. Any failures in the cell group ministry will be a failure in prayer. You should be committed to a time of intercession for your cell group and for the individual people in your group."

Daugherty adds, "We are determined to make our groups evangelistic. A group is not to be a holy club of believers that closes ranks and shuts others out. Each group must have a goal and directive, to feel that 'we're going to win the lost.'"

CHAPTER TWO: SMALL GROUP DISCUSSION

Icebreaker: Have you ever had a way of thinking or viewpoint that God has helped you to change? What impact did that have in your life?

Questions about this chapter:

1. Why did Daugherty start the first groups? Why do you think this attempt failed?
2. What did Daugherty do differently the second time he started groups? Describe the impact of this "second crop" of groups.
3. Why and how did Victory begin using "target cells"? How could target cells help most churches?
4. How did Victory's cells change as a result of Daugherty's ministry in Russia?
5. Discuss other changes Victory has made to enlarge or improve its cell system.

Application: Share one principle or practice you gleaned from this chapter on Victory's cell history. How could that be best integrated into your present or future cell group? Your church?

3

Blessed to be a Blessing

Victory's Creative Cell Groups, Part 1

Central to all Victory's cells is a strong focus on God's Word. Look in any Sunday bulletin, and you will find that week's cell lesson. Each cell lesson contains a title, introduction, simple outline with several Bible references, and discussion questions. Later lessons have a weekly Scripture memory verse and a "Be a Doer of the Word" application section. This focus on God's Word has prompted many cell members to report, "I finally feel like I understand what the Bible says."

Cell lesson material is based on Daugherty's past sermons or a book he authored on a Bible topic, and is written in a group friendly format. Cell leaders can either share the lesson printed in the bulletin, or lesson printed in any of six books, each filled with twelve months of weekly lesson outlines.

Daugherty himself is a skilled Bible preacher with a strong belief in prosperity and God's blessings. He teaches that God is a good heavenly Father who wants to bless his obedient children. Daugherty is also clear about prosperity's goals: God blesses us "so we can be a blessing to others."

"If you meditate on God's Word," Daugherty challenges, "and do what's right, you will prosper." Look at Matthew 6:33. "If you will seek

Sample cell lesson printed in Sunday's bulletin

VISION FOR VICTORY SERIES: #5 Ministry and Outreach

Purpose and fulfillment in life will come only as we do God's will. God's will is for every believer to be a minister of reconciliation. We minister in two ways: by the words we speak and by reaching out and touching others. We are to do the works of Jesus. We are His hands and His voice in this world.

I. We are all ministers
 A. Ephesians 4:12
 B. 2 Corinthians 5:18-19
II. We are commissioned to minister
 A. Matthew 28:18-20
 B. Mark 16:15-20
III. We are to do the works of Jesus
 A. John 14:12
 B. Luke 4:18-19
  1. Preach the Gospel to the poor
  2. Heal the brokenhearted
  3. Preach deliverance to the captives
  4. Bring recovery of sight to the blind
  5. Set at liberty those who are bruised
  6. Preach the acceptable year of the Lord

Let your light so shine!

DISCUSSION QUESTIONS

1. When did you discover you were a minister of reconciliation?
2. Share how you have applied Luke 4:18-19 to your life.
3. Share what Matthew 28:18-20 means to you.

Please e-mail, fax or mail any correspondence, attendance and praise reports.

first God's kingdom and His righteousness, all the other things you need will be added to you." Remember Philippians 4:19. "[Your] God will supply all your need according to his riches in glory by Christ Jesus."

Daugherty also stresses personal holiness, often referring to Psalm 1:1-3, "Blessed is the man that walketh not in the counsel of the ungodly, nor standeth in the way of sinners, nor sitteth in the seat of the scornful. But his delight is in the law of the Lord; and in his law doth he meditate day and night. And he shall be like a tree planted by the rivers of water, that bringeth forth his fruit in his season; his leaf also shall not wither; and whatsoever he doeth shall prosper."

Because of Daugherty's clear teaching, cell leaders see their groups and meetings as channels through which God can bless members. Those members, in turn, can bless others. One of the methods for blessing others is the outreach crusade designed monthly by Care Pastors.

Howard and Cheryl Shouse, Care Pastors over a three zip code district in Tulsa, shared one example of this. Their group of leaders banded together to do an outreach into the nearby River Parks and River Chase apartment complexes. They brought a yellow Kidz Club truck to entertain and share the Gospel with children, while they gave out lemonade, cookies, and 100 bags of food. As a result of that outreach, ten people made professions of faith in Jesus Christ.

Cells in the Pastoral Care Department

Victory's nearly 1,000 cells are organized under eight different departments and ministries, each wanting to bring as much blessing to their groups as possible. Key is Victory's Pastoral Care Department, which oversees three out of every five Victory cells, and is directed by the Popenhagens with the help of six "Care Pastors" and seven other pastors. This department coordinates Victory's cell leadership training and the apprenticeship outlined in chapter five, Victory's monthly "encounter weekends," and the joint monthly cell leader meetings with Pastor Daugherty.

DETAILED GROUP PROFILE
(January 2001)

TOTAL CELLS: 930
TOTAL ON SITE GROUPS: 382
TOTAL OFF SITE GROUPS: 548

Zip Code Cell Groups: 65 (off site)
Workplace Cells: 48 (off site)
Ladies' Cells: 43 (5/38)
Victory by Virtue Cells for Women: 25 (15/10)
Men's Cells: 8 (2/6)
Victory by Virtue Cells for Men: 17 (on site)
Young Single Adult Cells: 9 (2/7)
Singles' Cell Groups: 9 (2/7)
Young Marrieds' Cells: 16 (12/4)
International Cells: 27 (10/17)
Hispanic Cells: 19 (9/10)
Healing Cells: 6 (1/5)
Nursing Home Cells: 33 (off site)
Watchman Ministry Cells: 81 (1/80)
New Members' Cells: 2 (on site)
Prayer Fellowship Cells: 29 (on site)
Children's/Kidz Clubs: 166 (55/111)
Bus Ministry Cells: 35 (both on and off site)
S.O.U.L. Youth Outreach Cells: 14 (3/11)
Youth Cells (all types): 49 (13/36)
Sunday Cells: 33 (on site)

Continued on next page . . .

Continued ...

Growing Kids God's Way Cells: 4 (on site)
Nursery Cells: 35 (on site)
Marriage Cells: 5 (3/2)
Support Cell Groups: 3 (off site)
Family Ministry Groups: 6 (off site)
Music Cells: 22 (on site)
Sharon's Bible Study Cells: 20 (on site)
Senior Adult 65+ Groups: 3 (off site)
Prison Outreach Cells: 9 (off site)
Special Interest Cells: 17 (4/13)
Cyber Cells: 3 (off site)
VCS Cells: 3 (on site)
G-12 Staff-Directed Cells: 39 (both on and off site)
G-12 Area Coordinator Cells: 20 (off site)
Cell Leader Training Cells: 6 (on and off site)
Fellowship of Ministers: 1 (on site)

NOTE: Number after parentheses and to the left of the slash is for on site group/s; number to the right of the slash is off site group/s number

SOURCE: Pastoral Care Department Office

The Pastoral Care Department receives other departments' monthly updates on additions and changes in cell groups, and compiles this material to prepare the monthly "cell group directory." This directory gives Victory's complete listing of all current cells, leaders, meeting dates, times, and locations. It is one of the primary tools Victory uses to promote its groups. Lynn Popenhagen jokingly refers to tracking Victory's groups like tracking "the Dow Jones average, because our numbers change daily." This directory has up to 150 changes a month — leaders add apprentices, phone numbers change, new groups are added, and a few are deleted.

Each department submits a monthly cell report to the Pastoral Care Department, listing their cells' joint attendance and number of salvations, baptisms, healings, and new visitors. The Pastoral Care Department then combines these reports for an overall cell report.

Zip Code Cells

Foundational to Victory's cell system and the Pastoral Care Department are more than 60 "zip code cells," each welcoming participants from a geographical area, apartment complex, or neighborhood. Zip code cells are usually led by couples who hold weekly 90-minute home meetings that cover Daugherty's "five-fold vision" of worship, prayer, study of the

Word, fellowship, and outreach. Talk to Victory's Care Pastors, and they quickly recommend that a church establish zip code cells before diversifying.

Since 1983, these have continued to be Victory's "bedrock" cells. Near the end of each worship service, Daugherty invites those who want to receive ministry, or make a decision to accept Jesus as Lord, to come to the "altar area" in front of the platform. While all cell leaders are invited to do so, zip code cell leaders are most faithful to minister at the altar every worship service. In this way, they make direct contact with new people. Victory makes note of each Sunday visitor and altar respondent, and on Monday disperses the information to its staff and leaders. Zip code cell leaders thereby grow their groups in four ways: by meeting and ministering to new people at the altar; by following up on visitors and altar respondents in their areas; by welcoming people who see their listing in Victory's monthly "cell group directory;" and by inviting friends, acquaintances, relatives, and co-workers to their weekly meetings.

No matter which zip code leader I observed or spoke with, they were quick to share answered prayer. I visited Don and Susan Lipke's zip code cell meeting, held in a residential Tulsa neighborhood, and enjoyed a Tuesday evening filled with worship, the Word, and prayer. In the two years they had been leaders, the Lipkes saw a three-month-old baby healed of seizures, a drug addict delivered of addiction, an adulterous marriage restored, a father and caregiver born again in a nursing home, and many believers grow in the Lord.

Zip code cells have had profound impact on cell members, even helping in practical ways. Alan Pope told how a friend needed a new roof, so he called a member of his zip code cell for advice on roof repair: "Every able-bodied member of my cell group came to help put on a new roof. What does my group mean to me? Family."

Business Cells

Chief among other 'crown jewels' in Victory's cell system are their growing number of business and workplace groups. About 50 such "business cells" meet in the middle of company breaks, during lunch hours, in workshops, in factories, in boardrooms — whenever and wherever it is most convenient. Because of time constraints, some business cells only have 20-30 minutes for their weekly meeting, and either have a shortened version of the "five-fold vision," or focus on three of the five — prayer, the Word, and fellowship.

Ongoing prayer and care given to members during the week make up for the shorter cell meetings. Bruce Mow is a volunteer staff member and business owner who helps with Victory's workplace cells. Mow explains it this way: "If you are called to a business, and lead a business cell, that's your full-time ministry."

BENEFITS OF WORKPLACE CELL GROUPS

1. **Spiritual nourishment** for the Christians in that workplace, to aid them in their Christian walk.
2. **Platform for wide variety of ministry,** both during meetings as members minister to one another, and between meetings when seekers come for prayer and ministry.
3. **Creates evangelistic opportunities** with fellow employees who watch members' lives and often come with questions, comments, and needs.

Source: Bruce Mow
Victory's Cell Ministry Resource Book, p. 36.

Jim Husong would agree. Husong is a machinist whose shop makes components for utility trucks. Husong's workplace cell meets in his east Tulsa shop during Thursday lunch, involving salesmen to secretaries. Many who do not attend that cell have come to Husong and others asking for prayer for specific needs and concerns. One man, who now attends church with his wife, wrote, "Jim, you and your group have made such a positive impact in my life. My wife has noticed a big difference in my attitude and my language. She would like to personally thank you for having the heart and care to reach out to me."

STEPS TO START A WORKPLACE CELL GROUP

Prepare yourself: After you decide to start a workplace cell, make sure you have completed Victory's volunteer application and have a current background check on file. Go through the leaders' training and rally Christians around you who support your vision for a workplace cell.

Prepare your place and time: Secure permission from those in authority to allow the group to begin. Share the time and location with as many people as possible. Make sure the participants have received permission from their supervisors to attend meetings.

Develop a plan to raise up leaders and assistants to make sure all participant's needs are met: If appropriate, designate a praise and worship leader. Identify those who can lead in prayer. Inform all participants that you and your leaders are available for prayer at other times as well.

Determine your desired meeting flow: With the time available, decide time frames for praise and worship, prayer, testimonies, teaching, and fellowship. Be consistent and sensitive to work schedules in your start and stop times. Be sure you are familiar with each lesson, and prepared to flow as the Holy Spirit leads.

Source: Based on material by Bruce Mow, *Victory's Cell Ministry Resource Book*, pp. 36-7.

Another workplace cell in a Boeing plant so positively touched the workers they asked that group leader to become a chaplain at Boeing and start an afternoon training session for employees who were having marital problems. "That is an example," Daugherty stated, "of the corporate world asking the church for answers. A cell group opened them up to see that there was someone in their company who could help them do better."

Ladies' Cells

According to Lynn Popenhagen, Victory has had ladies' cells almost from the beginning. Many women feel safer with other women who understand what they are going through, and will open up more in a ladies' group than in a mixed group.

Victory's divisions for young single adults and older singles provide cells for single females who want to be with other singles their same age, while Victory's more than 40 "ladies' cells" reach out to

POSSIBLE TYPES OF LADIES' CELLS

Cells for seasons in a woman's life:
- Single
- Married
- New and expectant mom
- At home mothers
- Empty nesters
- Senior years

Support cells:
- Single moms
- Divorce recovery
- Crisis pregnancy
- Widows
- Weight issues
- Prodigal children

Special interests cells:
- Home-school moms
- Career and business cells
- Exercise cells
- Emotional healing cells

Source: *Victory's Cell Ministry Resource Book*, p. 32.

married and single women who enjoy more of a mixture. Some ladies' cells meet in homes in the morning or afternoon, times more convenient for full-time homemakers and female shift workers. Other ladies' cells meet in the evening in a home or apartment of a specific zip code area, drawing together those who prefer a group with other women, so they can share common interests and struggles.

Some ladies' cells focus on a special "season" in a woman's life, such as those 40 years and older, mothers with young children, or those with unexpected pregnancies. While most ladies' groups use Victory's weekly cell lesson, five groups have asked permission to use the book entitled, *Lady in Waiting*, a study of the book of Ruth geared to single women

LADIES' CELL TESTIMONY

A doctor told a mother in a "home-school moms' cell" that her son Jonathan had an abnormal pituitary and would have "very stunted growth." When the mothers in that ladies' cell prayed for that little boy, they were impressed with Luke 2:52, "Jesus increased in wisdom and stature, and in favor with God and man."

That mother focused her thoughts on Luke 2:52 during the six weeks she took Jonathan to different specialists. When she was finally able to see a Mayo Clinic specialist, he was almost rude. "Lady," he began, "I don't know why you're here. Your son does not have stunted growth. In fact, from previous test results six weeks ago to my test results, Jonathan has had the equivalent of a year's worth of growth."

Popenhagen remarked, "Every time we saw Jonathan after that, his pants always seemed too short for him. Praise God for home-school moms who know how to pray, get a word from God and be a strong support system for other moms!"

Source: Lynn Popenhagen

"waiting for their Boaz and not settling for Bozo."

Have these ladies' groups touched the lives of the women involved? Iva, Marie, and Jeri were recent widows who formed a weekly ladies' cell that provided them with the prayer and support system they needed to get through their grieving process; now they help other recently widowed ladies. When Geneva and her husband moved to Tulsa because of his job, Geneva grew lonely without female friends. That changed when she discovered a ladies' cell that met in her area: "Now in my ladies' cell, I have many friends who love me each week, and pray for me and my needs. Now I reach out to new ladies who come and feel lonely."

VICTORY BY VIRTUE CELLS

Victory by Virtue is a twelve-week course offered three times a year on Tuesday evenings. The first hour is for teaching and homework; the second hour women form more than 20 discussion cells that end with prayer and ministry. The final session of the course is a "virtue covenant ceremony banquet" where women come dressed in white, accompanied by sponsors who have supported them in prayer and encouragement.

Can Victory by Virtue discussion cells qualify as cell groups? Victory has two styles of cell groups: ongoing and short-term. While groups like the zip code, business and ladies' cells are ongoing, provision is made for short-term cells, like Victory by Virtue's 12-

VICTORY BY VIRTUE OUTLINE OF LESSONS

1. A Lifestyle of Virtue
2. You are Valuable
3. Motivational Gifts
4. The Study of Names
5. Purpose and Destiny
6. Juggling Your Priorities
7. Godly Communication
8. Avoiding Deception
9. Soul Ties
10. Single Women and Sexual Purity/ Married Women and Sexual Purity
11. A Victory by Virtue Woman
12. "Virtue Covenant Ceremony Celebration" Banquet

Source:
Victory by Virtue workbook

VICTORY BY VIRTUE CLASS FORMAT

Women in a large group:
- 15 minutes praise and worship
- 45 minutes teaching on that night's topic

Announcements and brief break

Prayer and discussion cells:
- 10 minutes reviewing that night's lesson
- 40 minutes responding to discussion questions and praying with one another

week prayer and discussion cells. Discussion cell leaders must first go through Victory's application process and cell training, and also serve as an apprentice. During these 50-minute cells, group members do at least three activities of the "five-fold vision": they pray, study and interact around God's Word, and fellowship together.

Are these prayer and discussion cells important? Nancy Mashburn, who works under Lynn Popenhagen to coordinate Victory by Virtue, puts it this way: "You can listen to a lesson, but that's not enough. A woman needs a place where she can personally share, pray, see God's Word in action, and get into ministry. A discussion cell is women ministering to women the love and grace of God."

I decided to go and see for myself. On a Tuesday evening, I attended a Victory by Virtue class on the topic of wrongful "soul-ties," held in the Victory Bible Institute chapel. That night nearly 100 ladies were present, and I spoke with as many as I could.

Most fascinating were these women's own stories. Leanna finished her Victory by Virtue course in the summer of 1997, went through training, served as an apprentice, and has since led a discussion cell. Leanna joyfully told how this course "completely" changed her life, overturning her former feminist attitudes. Her current apprentice, Kathryn, pointed to the importance of confidentiality, for "this is a special time for the ladies of our cell, and we want them to know we're here for them."

Connie told her story, repeated in essence by many women there: Her Christian husband left her and was gone for one month. During that time he prayed that God would give him a godly wife. Connie's discussion cell group prayed with her, because Connie could be controlling and nagging. "I learned I needed to yield to my husband," Connie explained, "and allow him to be the head of our marriage. When my husband returned, he was delighted. He now says I am the virtuous woman he always wanted."

Men's Cells

While many Victory men lead zip code and business cells, others prefer being in a group just with other men. Focal to Victory's men's cells is their monthly men's breakfast, with either Daugherty or a special guest speaking. Each men's breakfast concludes with an altar call, men then joining in groups of three or four for mutual sharing and prayer. This practice of forming small groups allows each man to get a taste for cells, and to want to continue in one.

There are three "varieties" of men's cells at Victory. Ken Weaver, staff pastor over the men's ministry, teaches a 12-week "Victory by Virtue" course for men, every class ending with prayer and discussion cells lasting 45 minutes. After graduation, men are encouraged to keep these groups going.

A second type meets in homes and restaurants, with men giving each other mutual support and ministry. Different departments at the church refer men who need ongoing ministry to the third type of men's groups, known as "deliverance and restoration cells." These on-site Monday and Thursday groups are filled with men who are alcoholics, addicted to drugs, struggling with crisis issues, or just need guidance in being a godly man.

While these last two types of men's cells are different, their format is similar. The leader first uses Victory's weekly lesson as a brief "devotional," then gives approved teaching specific to men. This additional material usually includes a section from Ed Cole's book, *Manhood 101,* or one of the 250 Ed Cole videos in Victory's lending library. After teaching, the cell leader will encourage each man to share what is happening in his life. Cell meetings usually end in prayer.

"Men struggle most with sexual sins," Weaver highlights. "Daily there are all kinds of opportunities for men to sin and lust. It's a moment by moment decision we men must make to stay close to God. That's why men need to be accountable to other men in a cell; that's the only way many will be able to stay free of that stuff. In a men's cell group they are able to bond with other men and see first hand what it means to 'walk with the Lord,' and receive unconditional love and acceptance, sometimes for the first time in their lives. That's why I consider men's groups the most important cells in a church."

Post-Encounter Cells

Victory encourages all members to participate in Encounter Weekends. These intense Friday night and Saturday daytime events, filled with material based on Daugherty's books *Breaking the Chains of Bondage* and *Turning Your Scars into Stars,* help the new and existing believer find fresh freedom from past sins, brokenness, and bondage.

AREAS OF CONCERN IN ENCOUNTER WEEKEND

- Fear
- Unforgiveness
- Sexual impurity
- Compulsive behaviors
- Religious and occult involvement
- Idolatry and rebellion
- Areas one was victimized
- Hurtful thoughts and habits

Victory began by offering three types of encounters three times a year: co-ed Encounter Weekends, so both husband and wife could go together; women's encounters; and men's encounters. Victory then discovered that people at encounters are more hesitant to share openly and to receive ministry when with those of the opposite sex. Starting in January of 2001,

Victory offered concurrent but separate monthly men's and women's encounters. Men and women join in praise and worship together, as well as eat side by side at mealtimes, but teaching, discussion, and ministry is done in separate men- or women-only encounters.

While Encounter Weekends have sometimes been held in church facilities, Victory found it is best to leave distractions in the city. Now most monthly encounters are held at Camp Victory. Set on Lake Keystone, Camp Victory is used year round for encounters, retreats, picnics, summer camp, outreach ministry camps, and missionary training. Activities range from swimming, basketball, fishing, canoeing, archery, hiking, water games, volleyball, go-carts, arts and crafts, and challenging group games. But few activities at that camp have so dramatically touched lives as Victory's Encounter Weekends.

Although Victory's Encounter Weekend format includes prayer and sharing groups, some find that an encounter is not enough. Leaders start 12-week "post-encounter cells" with a first meeting filled with icebreakers, so that participants will get to know each other. The remaining 11 weeks, trained leaders first go through the lesson in the weekly bulletin, then turn to a chapter in Daugherty's book, *Breaking the Chains of Bondage.* Some also study Sharon's book, *Avoiding Deception.* Both men's and women's post-encounter cells are listed under "men's cells" and "ladies' cells" in Victory's monthly cell directory.

Wendy Tyler had a special interest in young women going through difficulties, and though they were helped in Encounter Weekends, Tyler could still sense that more was needed. Tyler and others find that after an encounter, follow-through happens best in post-encounter cells. Post-encounter cells help members walk through any half-hearted decisions, and provide a group aimed at fellowship and trust, where men or women can freely express themselves.

YOUNG SINGLES' CELLS

Victory soon discovered that many singles did not want to be in cells with married couples. Currently a volunteer pastor oversees more than a dozen "young singles' cells" for those 18-28, and another staff pastor manages cells for singles 29-50. While both these pastors offer their singles separate weekend evening services, the greater focus is on their cell groups.

Young singles' cells target students from Tulsa's colleges and universities, as well as young working professionals. These groups meet in office buildings, homes and church facilities, with all but one monthly "outreach cell" doing Victory's weekly lesson and five-fold vision. Cell leaders aim at each young adult having a role in his cell.

Radhika Mittapali, mentioned already in chapter one, now not only manages investor relations in a multi-million dollar communications company, but also leads one of an increasing number of G-12 groups among Victory's young singles' cells. In her first group aimed at young single professionals, Mittapali got her friend and colleague Christina Grimm to join her, and by September of 1998 they led a weekly business luncheon. Their weekly cell group grew quickly, even sharing their message over the internet

HOW TO START YOUNG SINGLES' CELLS

Pray for God to bring you leaders with like vision (Prov. 29:18).

Establish approved teaching material to share (Col. 1:28).

Locate places to meet (Matt. 22:10), such as restaurants, homes, clubhouses, business rooms, church facilities.

Train your leaders in the areas they are called to teach and/or minister (Rom. 12:6).

Follow-up (I Tim. 4:6) by having a designated person or cell leader visit each group for support and accountability. This person reports back to the Singles' Pastor with updates, and suggests help needed by any group or groups.

Encourage (I Cor. 12:14), for we cannot fully operate without all parts of Christ's Body working effectively.

Source: Tracey Robertson
Victory's Cell Ministry Resource Book, p. 30-1.

to a growing e-mail list. Grimm, her former intern, now leads another weekly leadership G-12 group using Victory's lesson material. She also invites special speakers to their monthly outreach meetings, with an average of 30 business professionals in attendance, and more than 120 addresses on their e-mail distribution list. In her last e-mail note to me, Grimm excitedly wrote how two business professionals gave their lives to Jesus at a recent monthly meeting.

Tracy Robertson, formerly Victory's volunteer Young Single Adults Pastor, highlights the importance of answered prayer in their nearly ten target cell groups. The first cell Robertson led, they prayed that her friend Gordon would give his life to Jesus and get on fire for God, not that he would simply get off drugs. Within a year, Gordon had not only surrendered his life to Jesus and joined Robertson's cell group, but he was training to be a cell leader himself. Robertson now advises other churches reaching young adults 18-28, "First get a few young adults on fire for God. Then have them reach out in cells in their spheres of influence. You'll be amazed at the change they can make."

SINGLES' ADULT CELL GROUPS

Jodi Hill, pastor of singles 29-50, understands well the struggles and challenges mature singles face. Before she married her husband Dennis, she herself was a single parent with two sons, one with a disability.

Singles' adult cells meet in homes and church facilities, and start with a brief "nugget" or devotional from the lesson in Victory's Sunday bulletin, then launch into an approved topic for that meeting. These 15 singles' cells cover every topic from divorce recovery to weight control to in-depth Bible study, with focus on four types of singles: never married, widowed, divorced, and single parents.

"Many of our singles are already involved in zip code cells," Hill explained, "so we have more specific types of ministry. I constantly challenge singles to focus on God instead of their circumstances. Our cell groups provide the personal one on one ministry we all need."

YOUNG MARRIEDS AND "COUPLES' CONNECTION" CELLS

On a hot summer day in August of 1999, Steve Pogue was working in his garage, pondering the impact of Victory's thriving singles' cell ministry. Pogue, then a staff sergeant in the Marine reserves, had been a Victory member since his youth. "I realized," Pogue remembers, "that we needed something more for young couples at Victory. If couples, especially those ages 18-39, do not establish roots, they are very likely to leave. We wanted to create an environment where those couples could get to know each other."

So Pogue and his wife Wendy got together with good friends Bill and Michelle Cooley for prayer on Saturday nights, and later mapped out a written vision and mission statement. Busy young couples would more readily come to a bi-weekly fun and limited teaching event that gave necessary child care, followed by prayer and discussion cells. When Pogue approached Daugherty with his idea, Daugherty was delighted; he had already been praying for someone to help with the young couples. Daugherty then put the Pogues and Cooleys with Care Pastors Eric and Melody Castrellon, and they started with a Christmas Couples' Banquet that attracted 175 adults and

> **COUPLES' CONNECTION FORMAT**
>
> **The idea birthed in a garage** has taken definite shape. Each second and fourth Saturday nights from 6-7:00 P.M. the "Couples Connection Coffeehouse" meets in the back of Victory's west campus building, and provides free cappuccino, coffee, tea, lemonade, and occasional snacks to participating couples. Starting at seven, all gather upstairs for a joint time of prayer, announcements, testimonies, and a teaching "nugget" from someone in their growing leadership core. Then, while children are gathered in downstairs rooms in their own fun-filled cells, the adults form cell clusters for discussion, prayer and ministry. "This time," one couple told me, "is the highlight of our week."

more than 200 children.

Pogue reflected, "It's already been our joy to see several couples restored and many marriages improve. Not only do we have dozens of young marrieds in home cell groups and our bi-weekly cell clusters, at our Valentine's Banquet in 2000, we had 424 adults and even more children. God has blessed us."

Victory also has a growing number of home-based young married cell groups, first started in January of 1999. While most who go to the Couples' Connection have several young children, the majority of those in home-based groups are couples in their twenties who do not yet have children. Victory has used past banquets and marriage conferences to recruit these leaders, who are also in their twenties, yet without children. Potential leaders then go through Victory's cell leaders' training, with most groups ranging in size from three to five couples. In addition to teaching and discussing Victory's weekly lesson, they have a good deal of fellowship activity together.

International and Hispanic Cells

Daugherty's heart for the nations has made an impact on Victory's cells. Not only do many continue to "adopt" a Victory missionary, but in one young adult cell I visited, a young man proudly told how he had ministered during a missions trip to India, and he encouraged others to do the same.

Victory has also created cells to better minister to the "internationals" in Tulsa. The Pastoral Care Department's Hispanic pastor, Carmen Gil, oversees Victory's nearly 20 Hispanic cells, with one cell teaching Victory's cell training in Spanish, five meeting in church facilities, and the remainder in homes. Most of these cells apply Victory's "five-fold vision" and use that week's lesson translated into Spanish.

INTERNATIONAL CELL VISIT

On a warm summer night, I visited Victory's Friday night combined international cell service, and found myself in the midst of 11 nationalities. After visiting missionaries spoke, we watched a video from Indonesia, where Muslims had destroyed 400 churches in two years. Especially touching was the scene of 200 persecuted Christians living and sleeping in the same large room, now dubbed a "refugee center." With missionaries Ken and Cris Sandberg in the center, we prayed for God's protection and the fire of His Holy Spirit to surround them as they continued their work in the Muslim area of Mindanao in the Philippines.

Henry Bartlett placed his hands on Indonesia and southern Philippines, part of a large global mural on one wall, and prayed: "For every church destroyed in Indonesia, let seven more rise up to proclaim Your Word. We declare that Jesus is Lord over Indonesia and the southern part of the Philippines!"

This department's international pastors, Henry and Tonia Bartlett, oversee nearly 30 international cells — including Chinese, Filipino, French-speaking African, Indian, Japanese, Korean, Russian, Slavic, Thai, and Vietnamese language cells — each doing the five-fold vision in their own language. The Bartletts, an intercultural couple themselves, help in large Hispanic and Burmese "cell-ebrations," with a combined 300 in attendance, and another 420 involved weekly in their international cells.

Each month they hold a joint "pot-blessing," when cell members bring a wide array of delicious international foods. "Kindness means a lot to internationals," Bartlett has learned. "We have found that expressing God's patient kindness in our cell groups has drawn many to salvation."

What effect have these international cells had on individuals? Delann Dao, wife in an interracial Caucasian and Cambodian marriage, had gone to Victory since high school. A few years after her marriage to Huy, they started attending an international cell. "I felt immediately accepted," Delann began. "It was the first time I really felt a part of Victory. This cell has been a place I can give out to others, as well as receive."

HEALING CELLS

Almost 40 people volunteer to help in Victory's hospital ministry. During the year 2000, they made 4,071 ministry visits to 610 patients; of that number, they made 1,144 visits to 250 patients who were Victory members, and 2,927 visits to 360 patients who were non-members. When asked why they make more visits to non-members than to members, their answer is simple. People who regularly sit under faith-filled preaching of God's Word make less frequent trips to the hospital; when they do require hospitalization, their stays are not as long.

Most healing cell leaders come from these hospital ministry volunteers, with a typical four to ten people in attendance. About 90% of those attending healing cells are regular members, often composed of other hospital ministry volunteers; others are visitors, usually invited as a result of a hospital visit. There are exceptions, such as the healing cell that meets in Tulsa's Cancer Treatment Center; that cell has had up to 15 visitors. Bob Morton, overseer of Victory's hospital ministry and healing cells, meets with healing cell leaders in his G-12 group every

FAVORITE SCRIPTURES USED IN HEALING CELLS

"I was naked and you clothed me, I was sick and you visited me with help and ministering care" (Matthew 25:35a, Amplified).

"And He said unto them, 'Go ye into all the world, and preach the Gospel to every creature. He that believeth and is baptized shall be saved; but he that believeth not shall be damned. And these signs shall follow them that believe; In my name shall they cast out devils; they shall speak with new tongues; they shall take up serpents; and if they drink any deadly thing, it shall not hurt them; they shall lay hands on the sick, and they shall recover. So then after the Lord had spoken unto them, He was received up into heaven, and sat on the right hand of God. And they went forth, and preached every where, the Lord working with them, and confirming the word with signs following. Amen" (Mark 16:15-20).

HEALING CELL MEETING FORMAT

- **Begin** with praise and worship songs.
- **Pray** for others' needs, and that God would 'stir up' the gift of faith in all present.
- **Devotional** on Pastor Daugherty's weekly lesson.
- **Healing lesson** unique to the healing cells, with participants encouraged to share healing testimonies.
- **Lay hands** on any one who so desires, anointing him/her with oil.

Source: Bob Morton
Victory's Cell Ministry Resource Book, p. 52.

Wednesday, before Victory's evening service. During this time he gives a separate 'healing teaching' to augment the weekly lesson taught in the healing cells.

Healing cell leaders not only go through Victory's cell training, but also spend time in prayer and God's Word, and daily identify and declare healing scriptures. Morton explained, "I encourage healing cell leaders to be specific in their prayers and to live in a faith-filled attitude of prayer. That attitude is key, for when people are sick, they are more open to receive ministry than at any other time. We teach people it is God's will for them to be whole; some have never heard that before. As a result, we see many people saved and healed. Some healings are instantaneous, most are gradual."

Nursing Home Cells

Victory's nursing home cells began as Daugherty and his family went into nursing homes to minister. By 1985, Frank and Jane Johnson officially started Victory's ministry in nursing homes. In 1997, they turned oversight of the growing ministry to Jennifer Lamb, a geriatric nurse, and in 2000 leadership was taken by Terry and Glenda Mason. Victory's more than 30

NURSING HOME CELL TESTIMONY

Margaret had flown cargo planes during WWII, and in her older years had to be confined to a 'locked dementia unit' in a nursing home. The nursing home cell that met there for a year watched Margaret change. Like many of those residents who joined in those cell meetings, Margaret is no longer distraught and anxious, but relaxed, joyful, and smiling.

Source: Jennifer Lamb, R.N., M.S.N.

nursing home cells have two main formats and meet in nursing homes, adult day care centers, residential care homes, assisted living, and retirement centers.

One format is a meeting complete with song, teaching, prayer, and ministry. Another format is focused on one on one regular weekly or biweekly visits to particular residents, again having a time of song, teaching, and prayer. Most of these cells also serve communion.

Before starting a nursing home cell, a leader must contact the "activity director" at that specific nursing home for permission and scheduling. According to Lamb, the former director, these cells have three main benefits. First, leaders have led residents, even in their 90's, to receive Jesus into their hearts for the first time in their lives. These cells also provide needed ministry to elderly Christian residents. Thirdly, these cells help residents, who have time on their hands, to pray for family members and the spiritual life of the generations who will come after them.

Leadership encourages each zip code cell to "adopt" a nursing home, and visit residents when possible. This is crucial, because 60% of nursing home residents do not receive regular visits, and may only get family visits a few times a year — if at all. Lamb reported, "Through nursing home cells and 'adopt a nursing home' we have had testimonies of residents healed; many change from hopelessness to hope and joy. Repeatedly our cell leaders tell me they come away from a nursing home feeling more blessed than they thought possible."

Watchmen Cells

The fresh life in Victory's cell groups is greatly helped by Victory's follow-up to visitors and new members. Terry Glaze, New Members' Pastor, claims: "Follow-up is information conveyed with love. Love should be communicated in every piece of informational literature, every phone call, every warm handshake, and every radio and TV program. There are three keys to good follow-up: 'First, love them. Second, love them more. Third, love them exceedingly more.'"

Victory contacts new members, visitors, and altar respondents within 24 hours, quick to link them to a nearby or appropriate cell group. Members of a "Care Connection cell" call and visit new members and visitors the Monday after they visit and join. That same Monday Care Pastors make phone calls and drop-in visits, again pointing out the importance of belonging to a cell. Before the week is over, a nearby cell leader will also contact and invite a visitor to an upcoming cell meeting. This extensive system is tracked by software

VICTORY'S FOLLOW-UP PROCESS

Visitors and new members complete special cards. Care Pastors and cell leaders complete cards on altar respondents, including new believers. Once a person's card is received, usually on a Sunday, Victory has a set follow-up process:

Computer entry: Information from that card is entered in the computer by Monday at 4:00 P.M.

Primary distribution: This information is printed on individual/family follow-up referral sheets, and put in the PC Department's Mail Center by Monday at 4:00 P.M. for each applicable Care Pastor. The Care Pastor will contact each as soon as possible.

Secondary distribution: This information is also given to the appropriate department, such as a single mother's information to the Singles' Department.

Care Connection Cells: That Monday evening, "Care Connection Cells" meet to phone all those from Sunday's list, to make home visits to those within a three mile radius, and to write each a brief note. New members receive a letter hand signed by Pastor Daugherty.

Contact from Cell Leader: That week the individual or family will receive a call from a nearby cell leader, inviting him or them to the upcoming cell meeting.

New Members' Dinner: During this complimentary dinner for families and singles, new members can ask questions and receive personal prayer from Care Pastors, Watchmen, various staff, and Pastor Daugherty. They are encouraged to take Victory's Foundations Class.

Three month follow-up: Each week a list is printed for Care Pastors to review new members who have not yet become connected with a cell group during a three month period.

Source: Terry Glaze, *Victory's Cell Ministry Resource Book*, p. 61.

developed and tailored by Victory's New Member Department, and concludes with a three month follow-up check.

Glaze has not only worked hard on this new software, he is also a realist. While tracking new members, he found that even after repeated contact over a period of three months, some were still not connected to a cell group. So Victory formed "Watchman cells" to provide a minimal source of care and contact for these new members. Each ten "unconnected" new member families are assigned to a "Watchman," an Old Testament name for one who watches for the welfare of another. A Watchman makes at least one monthly caring phone call to each of his families, and quarterly plans a "get together" with them.

Currently there are more than 80 Watchmen cells. Some Watchmen make frequent home visits, and others invite target families to food and fellowship events. Watchmen also help people connect to other Victory cells. Some Watchmen have even gone on to become regular cell leaders, and some of their groups have become ongoing cells groups.

Are these cells in the Victory definition? Not in the strictest sense, for they do not meet weekly to apply even three activities in the five-fold vision. However, Watchman cells do provide minimum care, and are a "seed" that has sometimes sprouted into a full-grown cell group.

Is what the Watchmen do significant? Consider the words of Jerrylene Birchett, who with her husband has been a Watchman since 1997, and now assists Glaze in calling Watchmen overseers: "In a large church it's so easy for people to fall between the cracks. The same Watchman calling you for several months has much more impact than a series of people contacting you. We get to know people, their families and situations, and see them at church. Home visits are also good."

Birchett reflected, "I remember a Watchman named Nellie who called one of her people for months, and only got the answering machine. She always left a message and her phone number. One day that lady needed help, and called Nellie. People like that won't forget that we put out such effort to make them part of our church family."

CHAPTER THREE: SMALL GROUP DISCUSSION

Icebreaker: Share one specific way God has blessed you, and how that helped you then bless another person.

Questions about this chapter:

1. What impact has Victory's strong focus on God's Word made on their cell groups?
2. Why do you think Victory considers zip code cells to be "foundational"?
3. What impressed you most about the business cells? Why?
4. Consider the ladies' cells, Victory by Virtue discussion cells, men's cells and post-encounter cells. Which of these type cells do you think would be most helpful in your situation? Why?
5. Describe two differences you see between the various singles' and couples' cells. Which might be most workable for you? What was one interesting thing you learned about the international cells and healing cells? The Watchmen Cells?

Application: Share one principle or practice you gleaned from this chapter on the cells in Victory's Pastoral Care Department. How could that be best integrated into your present or future cell group? Your church?

4

FUELED BY PRAYER

Victory's Creative Cell Groups, Part 2

No matter which Victory cell — whether a zip code cell overseen by the Pastoral Care Department, a Kidz Club from the Christian Education and Outreach Department, or a high school campus cell connected with the Youth Department — each cell has a strong focus on prayer. Victory challenges all of its leaders to have daily devotional and prayer lives, to pray fervently before a group meeting, and to make prayer and ministry core activities in cell meetings.

Has all this prayer made any difference for Victory's cells? Remember the overall cell reports? In February of 2000 alone, Victory's groups reported 612 salvations. In the month of June, Victory's cells reported 549 salvations, 72 baptisms in the Holy Spirit, and 32 healings.

Daugherty explained Victory's view of prayer as he spoke to 100 pastors gathered for a workshop: "You can have a beautiful car, but unless that engine has fuel in it, you're not going anywhere. You can have a perfect model for your cell system, but unless you have the fuel of prayer, your groups aren't going anywhere."

"God gave man dominion," Daugherty stressed, "and Jesus commanded us to pray. If we don't take that dominion in prayer, the enemy rules by negative attitudes, by spirits of control, by various types of resistance. When you pray, especially when you reach a certain level of strong prayer, it clears the atmosphere of demonic control. Suddenly resistance is broken and people want to serve, to give, and to go to cell group meetings. Prayer is the fuel that allows God's blessings to flow in the church and in the cell."

Prayer Fellowship Cells

This emphasis on prayer is strongest in the nearly 30 "prayer fellowship cells" overseen by Victory's Prayer Department. Most have an hour-long meeting in one of Victory's prayer rooms that includes at least a five minute overview of that week's cell lesson, with the bulk of time given to worship and prayer. Leaders receive extensive training both through hands-on involvement in a current prayer cell, and through teaching and guidelines provided by this department, printed in Victory's Prayer Ministry Handbook; many also go through Victory's cell leadership training. Leaders call members between meetings; it is their responsibility to "pastor" their people.

Begun during the days of Victory's Bible Fellowships, these were first termed "prayer fellowships." When Victory changed the name to "cells,"

DISTINCT QUALIFICATIONS FOR A PRAYER FELLOWSHIP CELL LEADER

What are the unique qualifications of the leader of a prayer fellowship cell?

- **Willing to be a cell leader** with emphasis on prayer.
- **Knows how to pray** in the Spirit and with understanding (I Corinthians 14:15).
- **Willing to pray the answers** (not just the problems) according to Scripture.
- **Shows and models to others** how to pray.
- **Reaches out to others** and encourages them to join in prayer.
- **Assists a prayer cell leader** for a period of time.
- **Completes the cell leader training,** the prayer cell training, signs a commitment sheet and leads a cell according to the guidelines defined in the Prayer Ministry Handbook.

Source: *Victory's Cell Ministry Resource Book,* page 41.

they became "prayer fellowship cells," with greater stress placed on developing relationships with other like-minded people who want to pray.

While many prayer fellowship cells gather on a weekly basis and pray for a variety of issues, some spotlight specific concerns like missions, outreaches in other nations, and prayer requests sent in by missionaries. A few are termed "special prayer workshop cells," and include training on prayer in their two-hour long meetings. Before the 2000 presidential election, one woman started a prayer fellowship cell to pray for the election and for America; that cell continues to this day. Such initiative is encouraged, with the Prayer Department giving training, guidelines, and oversight.

Victory's Prayer Department, directed by Margaret Hawthorne, holds a weekly G-12 meeting with many of the primary leaders to give ongoing training and support. Hawthorne helps other departments in the church develop a prayer ministry or focus that fits them. For example, Hawthorne wrote the job description of a "prayer leader" now used in many zip code cells. She has also assisted with several

FORMAT FOR AN HOUR LONG PRAYER FELLOWSHIP CELL MEETING

What is the leader to do during these hour long meetings?

- **Start by introducing yourself** and acknowledging everyone present.
- **Worship together** in spoken word or song.
- **Share a five minute scripture teaching** based on that week's cell lesson. Give brief instruction on prayer, if needed.
- **Corporately pray in the Spirit and in English** as the Holy Spirit leads, encouraging the gifts of the Spirit to flow. Some prayer topics one may include:

  Monthly prayer topic sheet listing Victory's prayer concerns and needs.

  Victory's missionaries, outreaches, and schools posted for prayer.

- **Pray over prayer requests** placed in the basket in the room, careful to put them back for the next group who prays there.
- **Ask for prayer needs** of those present, and pray for their concerns.
- **Conclude** with prayer and thanksgiving.
- **Note:** Call each member during the week, especially first-time visitors. The leader is to "pastor" cell members.

Source: *Prayer Ministry Handbook,* pages 12 and 15.

MY VISIT TO HAWTHORNE'S G-12 PRAYER FELLOWSHIP CELL

By 2:00 P.M. on Wednesday, Margaret Hawthorne's G-12 leadership group of prayer leaders was in full motion. The 15 ladies gathered around the circle prayed over ten topics of concern for upcoming Word Explosion 2000, Victory's large annual conference, learned new ways to pray, and shared answers to prayer.

Besides regular prayer fellowship leaders, others gathered in that circle were: Lisa, a student who leads a Victory Bible Institute daily morning prayer cell; Esther, an intercessor for the Shouse's geographical district; Judy, who leads city-wide prayer for Tulsa's Hispanics; Chris, who is over the prayer chain that provides "prayer cover" for Daugherty when he travels overseas; and Shara, who helps oversee and connect others to "concerts of prayer."

Cosette, one of two intercessors for the singles' cells and ministry, told of a man who took nitroglycerin daily because of a heart problem; after he received prayer at a singles' Christmas function, he has not taken a single nitroglycerin tablet, and feels better than ever.

Prayer "S.W.A.T." (Spiritual Warrior Attack Teams), each with 5-7 people, continue to be contacted by phone to pray for the critically ill and for specific emergency prayer requests. They continue to join in prayer over the phone and in monthly S.W.A.T. cell meetings, where they concentrate on "breakthrough prayer" and quick mobilization.

Phyllis, who conducts prayer workshops, told of praying over the phone with a woman who had a sprained ankle. Her pain left immediately, and she could easily work and walk.

A report was given on "Convoy of Hope," when eighty-six churches in Tulsa combined efforts to feed the hungry and do outreach. Several prayer teams from Victory and other churches gathered throughout the day in the prayer tent to pray and provide a "prayer cover." There were no significant disturbances at what could have been a volatile event, and those who came after were surprised to find the grounds already cleaned.

I left that meeting sure of two things: God indeed answers prayer in Tulsa, and Victory is committed to prayer in ways few churches are.

prayer walks and prayer drives in various areas.

I asked Hawthorne why she thought Victory's cells placed such importance on prayer. She explained: "Our pastor takes prayer seriously. He gets up early every morning to pray, and he rarely misses his morning time with the Lord. On occasion he calls for the church to have corporate early morning or all night prayer. His attitude and focus on prayer has had great impact on our congregation."

According to Hawthorne, it was because of prayer fellowship cells that many more people "have gotten into prayer than would have otherwise. We learned that when a person doesn't know where he fits, we get him in a prayer cell, and then God will lead him, perhaps to another kind of cell. Others wanting to be intercessors find that a prayer cell allows them to fulfill what God has called them to do. We now have more than 1,000 people involved in Victory's prayer ministry. It is the cell concept that has helped bring this about."

CELLS IN THE CHRISTIAN EDUCATION AND OUTREACH DEPARTMENT

Prayer is also key to Victory's Christian Education and Outreach Department, which oversees one out of every five of Victory's cells. These are the most evangelistic of all Victory's cells, and are responsible for 70% of the salvations that take place through Victory's cells. The Christian Education and Outreach Department, with a primary spotlight on children, also reaches youth in the inner city, and has three varieties of cells: Kidz Clubs, bus route cells, and S.O.U.L. outreach youth cells. Because of its targeted focus, this department is one of the few that provides its own separate cell leadership training, with a focus on the importance of praying for one's children or inner city youth.

Rod and Gloria Baker, staff pastors who direct Victory's Christian Education and Outreach Department, are quick to point out that most churches only spend 4% of their entire budget on children, while 86% of all Christians receive the Lord before the age of 16. Rod Baker explains, "An effective children's ministry will reach entire families. If I were a betting man, I would put my money where I would get the greatest return. In the church, that would be on children."

Kidz Clubs

In 1998, when Rod Baker's son Jeremy came on Victory's staff to work with children, ages six to twelve, Jeremy thought cells were "just

MONTHLY SUPPLIES GIVEN TO KIDZ CLUB LEADERS

- **Faith-based curriculum** in a 'user friendly' format with a full planning schedule for meetings.
- **Props** and visual aids for object lessons.
- **Door hangers** and candy for visitation.
- **A different giveaway** (small toy) every week for each child.
- **Praise and worship song tape** styled for children.
- **Postcards,** including "We've Missed You," "We're So Glad You Came," and "Get Well Soon."
- **Laminated yard sign** with logo: "Victory Kidz Club meets here."
- **Bibles** and salvation tracts.

Source: *Victory's Cell Ministry Resource Book*, page 24.

something in the human body." But after hearing Daugherty repeatedly share his heart for cells, Jeremy and his wife Carissa finally understood their vitality and significance. This energetic young couple then rallied 26 people to do three-day "Backyard Bible Clubs" during that summer's Vacation Bible School (VBS) season. After these Backyard Bible Clubs ended, Baker challenged his leaders with questions, "Who's going to teach and disciple the kids you've reached? Who's going to win even more children to the Lord in our neighborhoods?"

Twenty-four of those 26 leaders then agreed to lead children's cells, called "Kidz Clubs," to help disciple the same children they had reached. Soon the Bakers developed "kid-friendly" material easy for any leader to use, complete with Bible-based lessons written just for ages six to twelve, and were busy challenging both Victory's congregation and the existing cell leaders with the importance of winning and discipling children. In addition to more than 50 facility-based children's groups, in a year and a half, the Bakers helped leaders start more than 140 off-site Kidz Clubs that meet every day of the week in homes, schools, apartment complexes, government housing, daycare facilities, and shelters for the battered and homeless. Since several Kidz Clubs are under different departments — such as the nearly 20 Kidz Clubs that meet in other rooms of host homes of zip code cells — only a few more than 160 children's groups and Kidz Clubs are listed under this department in the monthly directory.

Kidz Club leaders first go through an application and interview process, complete with a criminal background check. "We keep our training and materials so simple," Jeremy Baker explains, "that anyone can lead a Kidz Club. We make sure our leaders have all the tools they need to have a fun and interesting group. We use the same theme and lesson throughout our children's ministry — whether it be a regular Kidz Club or our five mobile Kidz Club trucks."

Baker trains existing and new Kidz Club leaders on the first Saturday of each month. At this meeting, Baker gives leaders an informational packet for each of that month's lessons. Each packet has a Word-based cell lesson designed for children, visual aids, toy giveaways, background music tapes, candy or snacks, Bible games, door hangers, and follow-up postcards. During this two hour session, Baker inspires his leaders with the vital importance of ministry to children. He then covers each of the lessons for that month, providing tips and practical insights to help each Kidz Club meeting be as fun and interesting as possible. The last week of the month, the lesson topic is always on missions; Victory teaches that each child is a missionary where he lives.

Kidz Clubs now rank first as the largest number of children's off-site cells in any church in the nation. Daugherty explains one reason for their success: "No one was really loving the children in many neighborhoods and apartment projects. So when one of our inspired leaders prepared cookies and punch and invited the neighborhood children to come for a 45-50 minute fast-paced meeting, those children flocked to that home or apartment. No matter who those children are, our leaders loved and ministered to them." Baker added: "Whenever a child comes into one of Victory's Kidz Clubs, he might feel like a 'nobody,' but by the time he leaves that cell meeting, it is our goal that he feel like a 'somebody.' God wants each child to be a champion for Him."

Bus Route Cells

Currently there are more than 30 bus route cells, overseen by Ed Brownfield, that bring 1,200-1,500 children and parents to Victory every Sunday. Six of these routes are to the handicapped and to inner-city low-income areas. Each bus captain, the leader of a bus route cell, must first apprentice as a "lieutenant" to a captain for three to six months, faithfully attending the weekly captain's cell meeting. Bus captains go through the same monthly training as Kidz Club leaders.

On Saturdays, most captains visit the homes of children likely to come on their routes the following day. This visit allows a captain to build relationships with the families of children who come on his bus and to invite youth and adults to go with their children to Victory.

According to Ed Brownfield, every Sunday each bus route cell has a three-part meeting. The first part, the captain shares the weekly cell lesson and interacts with children on the way to the church facilities. The second part, five and six year-olds divide into four classrooms, seven to eleven year-olds in three classrooms, teens go the S.O.U.L. meeting mentioned later, and adults attend the regular service. On-site children's cell leaders share the

BUS ROUTE CELL TESTIMONY

Brandon Lewis first came on a Victory bus at the age of 10 because he "liked the pizza." Brandon's bus captain nurtured him, and, by the time he was 16, Brandon became a bus captain himself. His bus route cell covers four large apartment complexes. He faithfully goes every Saturday to knock on doors, give out candy, and invite children to ride on his bus the next day. His Sundays are long, but Brandon doesn't mind.

"Each week," Brandon reports with delight, "I reach hundreds of boys and girls."

Brandon has since gone on a mission trip to Guatemala, with other trips planned. Now 19, Brandon declares, "Because one person, my bus captain, planted good seed, I am now reaching boys and girls around the world."

same children's cell lesson in their own unique style, further teaching that week's important Bible truth. This on-site meeting is comparable to an adult service geared to children, so the children will understand when they grow older and join in Victory's regular worship services. The third part of a bus route cell, the captain again ministers, this time as the bus takes children home. The captain then gives special prizes to riders, as part of recreational games that highlight the same spiritual concepts in a fun atmosphere.

These are "cells on the move," each bus route with 15 to 80 people, some larger routes adding a second bus. Les and Tammie Wallace have separately led two bus route cells as bus captains for two years. They start each Sunday at 8:00 A.M. with an inspirational bus captains' cell service, and by the time they finish their three-part bus route cells, are not done until 3:00 P.M. "It's worth it," Wallace stated. "Each Sunday we have about six to twelve children on our route receive the Lord. The kids out there need us. We're the only 'Bible' many of them will ever see."

S.O.U.L. Outreach Youth Cells

C. J. Jacobs remembers being a boy of nine, standing on the sidewalk to wait for a church bus to take him to Sunday morning and Wednesday night services. He wasn't sure what he enjoyed more, the Bible games or the peanut-flavored cookies the bus captain gave him. But it wasn't long before C.J. found his taste for cookies replaced by his newly found faith in Jesus Christ.

Now C.J. Jacobs and his wife have two teenage daughters, and the former stockbroker is Victory's Outreach Pastor, aggressively reaching inner city youth through S.O.U.L. outreach youth cells: "My personal focus is to reach teenagers, 13-20, in the Sunday service and mid-week cells channeled through our bus ministry. Most of their parents don't go to church. So we bring these teenagers in, teach them God's Word, and put them in mid-week cell groups that care for them on an ongoing basis."

The centerpiece of Victory's S.O.U.L. Outreach Youth Ministry, an acronym for "**S**tanding **O**n God's **U**nconditional **L**ove," is a fast-paced service held each Sunday morning, gathering 300 youth and workers, the majority picked up by Victory's bus ministry. In order to become a worker in this special service, you must go through an extensive interview, as well as a criminal background check.

Victory found that one Sunday service a week is not enough to minister to a teenager; S.O.U.L. cells were designed to reach these inner-city teenagers mid-week as well. Leaders for S.O.U.L. cells are chosen from the S.O.U.L. service workers who are willing to open their homes weekly or to find an available clubhouse. Leaders also go through three to five one-on-one training sessions with Jacobs in his office. This personalized training covers the purpose of cells, the importance of Victory's vision, and how to facilitate an interactive meeting. Each week, after the Sunday S.O.U.L. service, Jacobs meets in his G-12 group with his S.O.U.L. cell leaders and workers, giving them that week's tailored cell lesson and teaching them problem solving skills by using current situations and case studies.

S.O.U.L. cell leaders face the unique problem of transportation; not only do they provide a place and lead a meeting, they also have to provide transportation for the teenagers who want to come. This ministry is slowly acquiring passenger vans to help with mid-week transportation; in this way, even more S.O.U.L. cells can form.

Even with their unique challenges, S.O.U.L. cells now meet in volunteer homes and facilities, touching the lives of more than 100 teenagers. Five of the nearly 15 S.O.U.L. outreach youth cells meet in the Salvation Army shelter, the Lloyd E. Rader Youth Institute, one high school, and the David Moss Correctional Center. These all-male or all-female groups meet weekly for 90 minutes, sure to do three things: praise and worship; teach and interact; and pray and minister to each other.

Jacobs has "youthenized" Daugherty's weekly cell lessons for these inner-city, gender-based youth cells, including bits of material from "Women of Virtue" or "Men of Virtue." Leaders try to be creative with the youth, using fun icebreakers and occasional refreshments, as well as follow-up between meetings through caring phone calls to teenage members. "The cell group," explains Jacobs, "connects the heart, one person at a time, with the life of Jesus. We teach these teenagers that because of Jesus' power, they are not limited by their backgrounds or circumstances."

CELLS IN TH YOUTH DEPARTMENT

While the S.O.U.L. outreach youth cells touch Tulsa's inner-city youth, four types of nearly 50 cells in the Youth Department focus efforts on many of the other 50,000 teenagers in Tulsa. Tom Dillingham, staff pastor who directs this department, had formerly been a youth pastor of a church in Missouri, growing their youth group from 20 to 500 in six years.

When Dillingham came on staff at Victory, there were not many cells operating in the Youth Department, and it has been a slow building process. Although Dillingham did not have cells in his former youth group, he is now a vocal cell advocate. "Every time I examine the weaknesses of the former youth group I led," Dillingham comments, "I find that they could have been solved by cell groups. Cells solve the weaknesses of little or no real relationship, of discipleship, and of accountability."

Leaders first go through "Basic Training," a ten-week course designed specifically for youth cell leaders. Taught by Jim DePriest, who oversees the youth cells, "Basic Training" covers both doctrinal and practical topics essential to youth cells. Training subjects include: the Word of God in cells; prayer in cells; how to do prayer and worship in cells; evangelism; the purpose of cells; your legal rights, including how to set up a campus cell and find a sponsor; marketing and advertising ideas; ideas for increasing group fellowship and relationships; commitment to Victory.

According to DePriest, the weekly adult cell lesson did not always relate well to youth. Sarah Daugherty, who serves as junior high pastor, now writes the weekly youth cell lesson, which might be on the same topic or texts as the adult's, but is tailored to relate to teenagers. DePriest, who has coordinated the youth cells under Dillingham's direction since February of 2000, sends weekly e-mails to his leaders, and tries to call each one every week or two. Once a month, DePriest invites all youth leaders out for a time of fun and fellowship, whether that involves paint ball, laser tag, or "rappelling," rope climbing down a sheer cliff.

Ongoing leadership training is also done throughout the Youth Department. On Monday night, Dillingham meets with his G-12 leadership group of college students who help him oversee the youth cells and ministries, encompassing about 750 teenagers. Most cells are youth led, with their evangelistic efforts aimed at developing relationships with other teenagers, and then inviting them to Wednesday and Sunday night "24-7" youth services. These 24-7 youth services are complete with strobe lights and a youth worship band sure to rock any teenager's world. While some youth cells have had young people receive Jesus during their meetings, most find that cells prepare teenagers' hearts to accept Jesus as Lord in a 24-7 youth service. The Youth Department has a goal of increasing to 100 youth cells by the end of the 2002 school year.

School Campus Cells

More than half of these youth cells meet on 12 junior and senior high school campuses around Tulsa. Two schools even have multiple campus cells led by different leaders, overseen by a single teenager. This is part of a growing trend. These teens know their legal rights, are trained to find a teacher-sponsor, and lead a weekly 30-60 minute group that covers Victory's five-fold vision, complete with a youth cell lesson.

> **STORIES OF YOUTH CELLS ON SCHOOL CAMPUSES**
>
> **A seventh grade girl** in Jenks Middle School started a cell with 18 fellow students at her first meeting; many have already become born again Christians.
>
> **Dave** started a youth cell in Webster High School that grew to an unusual number, showing the potential impact this kind of cell can make on a school campus. Forty teenagers attended his first cell meeting. At his second meeting, 80 came; at their third meeting, 230 came and 18 gave their lives to the Lord. The group became so large that it had to meet outside.
>
> **Source:** Tom Dillingham

Youth task cells

Included in the nearly 50 youth cells are more than a dozen on-site "task cells." These task cells meet 30-60 minutes before one of the weekly 24-7 youth services in a spacious room on Victory's east campus. They consist of the "junior high altar workers cell," "senior high follow-up cell," and others. During this time, they briefly cover the weekly youth cell lesson, then pray and minister to one another. This allows a task cell to become a "team" with caring relationships, not just a disjointed group of teenagers with a job to do.

Youth target cells

Nearly a dozen cells target a specific age or gender group, and meet in church facilities and homes. These 60-90 minute weekly meetings cover the youth cell lesson with a special focus on discipleship. Groups like the one for junior high boys also cover material specific to the young men's needs.

"House Parties"

Victory sends some buses as far as 30 miles to bring youth to the Wednesday and Sunday night services. "House parties" are centrally located homes where youth meet 30-45 minutes before the buses arrive. House party "parents" are carefully screened and selected, and lead a brief Bible study on that week's cell lesson, then have prayer and ministry with those gathered.

Cells in the Sunday School Department

When Daugherty decided that each department should have its own cells, Victory already had topical Sunday School classes for adults and age-graded classes for children. But Gary Stanislawski, who has served as volunteer director of the Sunday School Department since 1995, knew the classes in his department would have to make a major transition to what are now known as "Sunday cells." Stanislawski, formerly an Air Force pilot and now an ordained minister with a master's degree in Christian education, was determined to join in unity with Victory's larger cell group vision.

According to Stanislawski, helping Sunday school teachers make the complete change to cells is a continuing process. Because Victory had groups since 1983, many already understood the concept; the hard part was helping them see themselves as cell leaders who minister and encourage mutual ministry, not just as teachers in a classroom.

VICTORY'S SUNDAY CELLS FOR THE YOUNG

Nursery & Children at 8:30 A.M.

- 6 weeks to 2 years
- 3-4 years
- K-5 years
- 1st Grade
- 2nd Grade
- 3rd Grade
- 4th Grade
- 5th-6th Grades

Youth at 8:30 A.M.

- 7th & 8th Grade Jr. Hi.
- 9th-12th Grade Sr. Hi.

Children's Church at 10:00 A.M.

- Nursery (6 wks. to 3 years)
- 4 year olds
- K-5
- Grades 1st through 6th

SUNDAY CELL TESTIMONIES

One mother with several small children enrolled in a new "Growing Kids God's Way" Sunday cell, even though she and her husband were separated. Near the start of the 18 weeks, her husband agreed to attend the meeting that focused on "The Father's Mandate." During that meeting, he heard about God's perspective on the importance of the father's role, and his heart changed. He attended the rest of that Sunday cell with his wife. They are now happily reconciled and involved in Victory.

An adult with severe mental and physical limitations attended the Sunday cell "Learning unLimited," a special needs cell for those with mental and physical challenges. She had not spoken for two years. Then, during praise and worship, as cell leader Shirley McCoy was speaking and touching each member, she responded aloud, "I love you, too." Medical care-givers were amazed that verbal skills began developing in such a severe, long-term case.

Source: Gary Stanislawski

Now, all adult Sunday cell leaders, still called "teachers" by many, are encouraged to go through Victory's cell leadership training. Each year, these leaders have a one-day annual retreat. The theme in 2000 was "Releasing the Anointing," with an emphasis on ministering to members and encouraging them to minister to each other. Every six weeks, Stanislawski has a modified G-12 group with his adult, youth, children's, and nursery coordinators, as well as Barbra Smith, a full-time department secretary. Every week at 7:59 A.M., he meets with all Sunday cell leaders for prayer before the 8:30 cell classes begin.

More than 30 Sunday cell classes meet in three buildings at 8:30, 9:00, and 11:00 A.M., and cover at least three parts of Victory's five-fold vision of worship, prayer, teaching God's Word, discussion, and fellowship, as well as mutual ministry with outreach. Sunday cells range from Victory's nursery cells, which include infants six weeks to two years, to Sunday cells with topics for singles, for those married with family, to Keenagers, which aim at senior adults 55 years or more. Adult Sunday cell classes use their own lesson material. Topics include such things as "Growing Kids God's Way," finances, overcoming daily obstacles, in-depth Bible study, and the Jewish roots of Christianity. Each quarter, to further develop relationships, each adult cell class has a

"Sunday Morning Mixer," where they bring food to Victory's cafeteria for a time of fellowship from 8:30 to 9:00 A.M., then return to their classroom for an abbreviated meeting. Sunday cell leaders make calls in order to get to know their student-members better, and they either call or send a postcard to missing members.

While traditional Sunday school tends to be an information exchange from teacher to learner, 'Sunday cells' are more dynamic, an interactive approach where the teacher is a facilitator-leader who knows that members bring a wealth of life experiences to share in the process. So how does one help the traditional Sunday School teacher make the shift to interactive on-site 'Sunday cells'? Stanislawski recommends three principles: "First, the teacher-leader must learn to be led by God's Spirit, quick to have interaction and ministry in the classroom setting. I tell them not to be boxed in by a lesson plan. Secondly, they must be sensitive to the needs of people in their cell, know what they like and don't like, and find out about their families and how their lives are going. They must make sure each meeting includes a healthy time of prayer and ministry to one another. Thirdly, every 'student' must be prepared to be an active learner, willing to participate in ministry to one another."

Has this transition to Sunday cells taken hold? To a great extent, yes. A few leaders still consider themselves just teachers in a classroom setting, but many of these "cell classes" now have a fresh, interactive dynamic with a greater stress on developing relationships. Already some assistant teacher-leaders have started new Sunday cells, and nearly 30% of Sunday cell teacher-leaders also lead off-site groups. Victory has found that Sunday cells are a good way to train more leaders, some of these even launch into full-time ministry. Such was the case of former Sunday cell leader Brian Ruggles, now leader of the second-year program at Victory Bible Institute (VBI), and Fred and Janice Rivers, now associate pastors of a church in Alabama.

Cells in the Personal and Family Life Ministry Department

Victory began its Personal and Family Life Ministry Department in the mid 1980's to provide pastoral counseling for members of church families. Started by Charles and Margaret Hodge, this department is now headed by Harry Latham, and has three full-time and one part-time staff members, aided by a group of trained volunteers and graduate students who desire more experience in Christian counseling. This department tries to respond to all requests for counseling, whether or not they are from Victory, and averages 50 counseling appointments per week, in addition to responding to letters, e-mail notes, and telephone calls.

The Personal and Family Life Ministry Department serves as a prime resource for all Victory's cells. Leaders are encouraged to refer troubled members to this department for individual counseling. But that is not the end of this department's cell involvement. It also ministers through the channel of three types of nearly 20 weekly cell groups: marriage cells, support cells, and family ministry cells.

These cell leaders take Victory's cell leadership training. In addition to which, all department volunteers, including the leaders of its marriage and support cells, must go through eight weeks of departmental training. This introductory orientation to Christian counseling, taught by Nikki Latham during the Sunday cell hour, covers biblical responses to basic counseling situations. Leaders for the marriage and support cell groups must also serve six full months as apprentices before becoming cell leaders.

Leaders receive ongoing support in two ways. First, this department has a monthly motivational meeting for all its leaders. According to Latham, the main objectives of this meeting are to appreciate and to encourage their committed leaders. Second, these leaders join the rest of Victory's cell leaders in their monthly meeting with Pastor Daugherty.

Marriage Cells

Each marriage cell involves about seven couples. These home meetings last approximately 90 minutes, with the weekly cell lesson applied to marriage, and often supplemented by a Christian video on marriage. All additional materials must be first approved by the Personal and Family Ministry Department.

Harry and Nikki Latham's personal journey best typifies the impact of Victory's increasing number of marriage cells. Divorced and searching for a closer walk with God, Nikki moved to Tulsa in 1989; she went to one of Victory's singles' cells. In that cell, she became

MARRIAGE CELLS

Marriage home-based cells: When focused on young married couples with children, it is important they have a simultaneous Kidz Club. After a nugget from the weekly lesson and its application to marriage, the group watches video tapes, follows the Couples' Devotional Bible, or uses other resources. Whatever the focus of this group, it is vital they lift up the scripture keys of fidelity, finances, prayer, and healthy role definition.

Preparation for marriage Sunday cell: "Before You Say I Do" is an instructive class complete with video tapes and a workbook covering multiple marriage topics. It is the first time many people will connect with other Christians as a couple.

Specialty groups: These marriage and enrichment groups provide teaching and mentoring for newlyweds, blended families, or whatever season of marriage is relevant — using appropriate resources.

Sunday marriage cell: "Triumphant in Marriage" is a Spirit-led and Scripture-based marriage study and support group.

Marriage skills training: An in-depth study and mentoring for couples, with stress laid on communication skills, using a workbook and Christian study guide.

Restoration Cells: These marriage cells focus on emotional healing, soul restoration, and healing from co-dependency.

Source: Harry and Nikki Latham
Victory's Cell Ministry Resource Book, pp. 45-6.

grounded in God's Word and grew in relationship with new friends. Like many in her group, she also became a Bible school student and short-term missionary.

While stationed at a table in the foyer to sell tickets for a singles' event, Nikki met Harry Latham, who was also divorced. Soon the couple talked of their love for missions and grew in their love for each other. They attended the marriage preparatory Sunday cell, "Before You Say I Do," a cyclical ten-week class, complete with a workbook, that covered topics as diverse as communication, finances, sex, and in-laws. "We learned so much in that Sunday cell," Harry remembers. "After the first class, Nikki and I just looked at each other saying that if we'd known before what we knew now, neither of us would have had to suffer divorce. We are so glad God forgives and restores."

LOVING THE UNLOVELY: MARRIAGE CELL TESTIMONY

A talkative lady attended a marriage Sunday cell, and Nikki Latham felt she should invite the woman to their home-based marriage cell. The woman gladly came, but she didn't have a car and needed Nikki to pick her up and drive her home. When she came, she monopolized meetings as she talked about her viewpoints and sorrows.

Four months later, in the middle of the night, Nikki's husband Harry woke to the phone's ring. An emergency room nurse called to say that the woman was dead, and that they found Nikki's name and phone number among her belongings. Harry found the husband and took him to the emergency room. Cell group members were her only 'family' at the funeral.

"I finally understood why we had to have her in our group," Nikki related. "For the last four months of her life, we were the only real family she had, pointing her to Jesus. She's in heaven now, with all her needs met. I picture her there, talking and talking, as Jesus draws her closer to Himself, 'Come here, daughter. Tell Me more.'"

Source: Harry and Nikki Latham

After they married, Harry and Nikki shifted to the marriage Sunday cell, now called "Triumphant Marriage," and soon made friends with other Christ-centered married couples. Harry and Nikki went through the cell leaders' training and later started the first of their own home-based marriage cells. After praise and worship, they shared the weekly lesson and discussed how it applied to marriage. Like other marriage cell leaders, they used resources like the "Couples' Devotional Bible" and the video series "Biblical Portrait of a Marriage" to trigger discussion and ministry.

"Several couples with problems in that group got back together," report Harry and Nikki. "One dating couple made the quality decision not to marry and some newlyweds got solid building blocks for marriage. We all began to reach out more to hurting people as we became more aware of the marital resources we did have."

The Lathams' favorite moment came when an usually quiet husband shared a choice nugget. The husband told what a difference 'cuddling' made in their relationship; he said if they cuddled for fifteen minutes in the morning, the entire day was better. Even a year after he shared that practical insight, couples were still helped by it.

Support Cells

Why are support cells important? Because, explains Charles Hodge, "Some men and women feel dirty, defiled, unloved, discarded, rejected, and without hope. People in their lives have misused and abused them. Many in our support cells experience a new beginning and are now living an authentic Christian lifestyle."

Nadina Stevenson, a Ph.D. and former Presbyterian pastor who now serves as a pastoral counselor in this department, works with the support groups. Some call these "restoration cells," for our Divine Shepherd "restores my soul" (Ps. 23:3a). During support cell meetings, participants first focus on God as they pray and use the weekly lesson as a devotional. They

SUPPORT CELL TESTIMONY

Dr. Hodge led a group of men through a cell focused on sexual purity. Even though he was a Christian, one man's past of perversion, unfaithfulness, and deceit affected his present. During that group, he finally came to genuine godly sorrow, and he fully repented to his wife and children, his pastor, and his counselor.

That man became fully accountable to this wife and counselor, and regained their trust. He is now in full time pastoral ministry, with his forgiving, faithful, and trusting wife by his side.

Source: Charles Hodge

then "work on themselves" as they learn and discuss a specific issue like anger, fear, codependency, compulsive behavior, or depression. Their goal is to apply biblical truths in order to experience victory and restoration in their areas of former defeat. Meetings conclude as members share prayer requests and pray for one another.

When a person's problem is more than a group or counselor can handle, Stevenson refers that person to Rapha Christian counselors or a domestic violence intervention counselor, and he helps that person keep a link to Victory. "Sometimes we're like a corked bottle," Stevenson illustrates. "We might have wonderful spiritual insights and deposits inside us, but nothing gets out because of the 'cork' of excess baggage — being raised in a dysfunctional family or problems in the soulish realm of our minds, wills, and emotions. Support groups help deal with those problem issues and 'uncork' a person so he can be restored to who God intended him to be."

Family Ministry Cells

Charles Hodge, previous director of this department, was concerned that many of Victory's families were so busy they could not take the time to be in any of the cell groups. Not only that, they were not even taking time to gather as a family for devotions and prayer. Hodge, a member of the American Association of Christian Counselors, and his wife Margaret knew something must be done to strengthen these busy families. So he began talking to a few husbands, challenging them to take an hour for weekly devotions with their families. These husbands would apply at least three parts of the five-fold vision and teach the lesson printed in the weekly bulletin to their families. Soon busy, childless couples and busy, large families held regular Family Ministry Cell meetings.

Hodge admits that he tried to get this started throughout the congregation, but that there are still just a handful of these family ministry cells. He credits the low numbers to a lack of promotion, and not having a system of accountability with these husband-leaders. These are the only type of Victory cells that do not require any leader training, with only a limited amount of screening. Husbands, or the head of any household, simply gather family members for a weekly, tailored "cell devotional."

Even with their small numbers, have Family Ministry Cells made any difference? Henry and Tonia Barlett, who head Victory's international cells, found that their children began to open their hearts during their family cell meetings. Their children excitedly told neighborhood children, who then asked to attend. That cell grew even more as these children persuaded their parents to attend, and many came to the Lord.

Cells in the Music Department

Victory links prayer with worship, key elements in their on-site music cells. Sharon Daugherty serves as worship pastor, careful to place each person in one of Victory's more than 20 music cells. This department chooses music cell leaders on the basis of their pastoral disposition and musical gifting. A musically talented person who does not have much interest in pastoring his cell members is paired with a co-leader who might not be as musically gifted but will help meet pastoral needs. If a leader is both musically talented and has a pastoral focus, that music cell will have one leader; if not, there are two co-leaders for a given cell.

Most cell leaders first go through Victory's cell training. Each music cell contains no more than 12 people, who meet for half an hour before practice. During this half hour, the cell leader not only covers the weekly lesson from the bulletin, but also encourages members to interact, pray, and have mutual ministry. These groups provide both accountability and a sense of belonging.

MUSIC CELL GROUPS SCHEDULE

There are a variety of music cells, coordinated by Victory's Music Director, Tim Waters:

- Choir cells meet Thursday evenings, 30 minutes before rehearsal.
- Male musicians meet from 12:30 to 1:30 on Thursday afternoons.
- Orchestra meets on Saturday morning with cell incorporated into rehearsal time.
- Student Worship Team Leaders meet Thursday morning during the "School of Worship."
- Music teachers meet every Friday morning for curriculum planning, student evaluations, and prayer.
- Front Vocal Team meets on alternating Saturdays with cell incorporated into rehearsal time.
- Instrumental cells meet on alternating Saturdays, overseen by co-leaders for each instrument (drums, guitars, etc.) and include prayer, instruction, and accountability.

Source: Miriam Springer
Victory's Cell Ministry Resource Book, page 53.

There are some specialty music cells, like those for drums, guitar, and keyboard. Before weekly Thursday evening worship rehearsal, the worship team and choir are divided into cell groups. From 6:45 until 7:15 clusters of choir cells (called "care groups" by most) meet around the church sanctuary, and spend their meeting praying and ministering to each other. According to Miriam Springer, Victory's Music Minister, "These cells have knit the choir and worship team together. When we stand before the congregation on Sunday morning, we no longer perform; we worship and lead others into worship."

Since many in the choir music cells are also members of zip code and other cells, they are no longer called music cells, but referred to as "care groups," with the cell leader being called a "choir care pastor." Richard Jaeger serves as "membership coordinator" and helps integrate new members into the choir, as well as oversee more than ten "choir care pastors." Choir groups usually have eight to ten members, with composition determined by zip code and gender. Sharon Daugherty adds, "It is important that we all stay connected. The enemy likes to isolate and pull people away from others so he can play with their minds. God's will is shown in Ephesians 4:16, that we instead be 'fitly joined together.'"

Tim Waters, Director of Victory's Music Ministries and VBI School of Worship, points out that evangelism has been the weakest aspect of the music cells. To strengthen evangelism in the music department, they are currently developing self-contained worship teams, which would be task-oriented "music cells." Each team would have a main worship leader, a pastoral person who would function as the "cell leader," a coordinator (to handle schedules, logistics, etc.), and a specified sound person. How do these help increase evangelism? The all-male worship team focuses not just on the men's ministry, but also on outreach and ministry to men's prisons; the all-female worship team helps not just with Pastor Sharon's women's meetings, but also ministers in women's prisons. Other worship teams, such as the primarily African-American worship team, assist in outreach events at the new "Dream Center," built in Tulsa's inner city to reach out to the less fortunate.

Waters finds even the current music cells of great benefit. They help 'pastor' his people, and they offer a vital leadership outlet to some who may never sing a solo or become strong musically. Music cells can also be used to initiate an area of innovation, such as a team of in-house songwriters. Waters comments, "For the young or new music program, I suggest starting with the music cells first. This not only builds a strong spiritual foundation early, it also builds unity among the team. Issues can be dealt with and aired there, rather than in the heat of a

MUSIC CELL TESTIMONY

A lady from Nigeria had miscarried her first child just before she started attending Victory. During her next pregnancy, she was in Victory, serving in the choir, and doctors told her this fetus also had severe problems and was traumatized.

Her choir cell leader told other leaders, and soon several prayer warriors were interceding for her with the entire choir corporately praying on her behalf.

She went full term with that baby and gratefully gave birth to a beautiful boy. When she stood before the choir to share her story, she commented most on the choir's faithfulness to pray every week until she had her "breakthrough."

Source: Miriam Springer

service. As the music program grows, the foundation of music cells helps keep negatives in check, such as the spirit of competition or offense over personnel selection."

Pastor Sharon's Bible Study

Each Thursday at 12:15 P.M., women in VBI's chapel sit behind wooden tables spread with Bibles and notes, gathered for a mid-day Bible study with Sharon Daugherty. Overseen by Pam Cornwell, volunteer director of Victory's women's ministry, this Bible study is a vital link in the lives of the ladies who participate.

Sharon Daugherty, honored by the congregation as a co-pastor with her husband, has held this mid-day ladies' Bible study since 1988. Through the years, Sharon has taught on a wide range of subjects: books of the Bible, like Romans, Ephesians, I John; and topical studies, such as "The Joy of the Lord," "Overcoming Strongholds," and "Developing Healthy Relationships." At times, Sharon invites other female staff to teach particular subjects. At other points, she has had Cornwell teach Bible studies.

When Sharon or others speak on a series, the women gather in small groups, called "study cells," after the teaching. These groups are similar to the prayer and discussion cells in Victory By Virtue. Potential leaders are hand-picked, then they apprentice with Cornwell and other leaders. Study cell leaders join together for a full hour before the Bible study begins. After teaching, each leader gathers with the same ladies to cover prepared questions and a discussion guide on that day's topic.

I visited Pastor Sharon's Bible study on a summer Thursday. Pam Cornwell taught basic and in-depth Bible study skills through a verse-by-verse study of the Gospel of John, part of a much longer series. After she taught for about 50 minutes, I watched nearly 100 ladies form study cell groups, with six to eighteen women in each cell. Each group spent approximately 45 minutes in review, discussion, questions and answers, and prayer.

What difference do these study cells make? Consider what four women said. Sylvia: "This group is a big blessing to my life. We all want to know and do God's will. In our cell we get answers to the questions we ask and insight into God's desires. Not only that, we also have the opportunity to share with each other."

Joy: "When I'm in our study cell, it's like a picture coming to life. This group makes God's Word real for me. Without it I would find the Bible intimidating and confusing."

Christy: "I look forward to my group. In a large church it's our opportunity to get together and know each other, and to share prayer requests."

Rebecca was most pointed: "This group is the 'glue' to my faith for the week. It helps to get me through to Sunday."

CHAPTER FOUR: SMALL GROUP DISCUSSION

Icebreaker: Briefly share one answer to prayer you have experienced. How did that affect your life? Your relationship with God?

Questions about this chapter:

1. Victory's focus on prayer is felt in all their cell groups but is best expressed in their prayer fellowship cells. How might a greater focus on prayer affect your current or future cell(s)?
2. Consider the cell groups in the Christian Education and Outreach Department, as well as in the Youth Department. In which one type of cell in these departments would you most like to be a leader or member? Why?
3. How are the Sunday cells and the cells in the Personal and Family Life Ministry Department similar? How are they different?
4. What interested you most about Pastor Sharon's Bible study discussion cells or about the cells in the music department? Why?
5. Which of the cells described in this chapter did you find most creative? Which do you feel has potential for possible use in your situation?

Application: Share one principle or practice you gleaned from this chapter on many of Victory's diverse cells. How could that be best integrated into your present or future cell group? Your church?

5

For Such A Time As This

Victory's Leadership Training

From 1983 until October of 2000, more than 4,000 people have gone through Victory's training and started cell groups. Daugherty is direct on this topic: "We will not launch you as a leader without proper training. You must first go through the Foundations Class, the cell training, and serve as an apprentice in an existing group. When the staff pastor and cell leader you work with feel you are ready to start your group, then we will release you to start your own cell."

After Victory's first failed attempt at groups in 1979-80, Daugherty made training a major ingredient in Victory's cell success. Before Daugherty re-started groups in 1983, he spent nearly nine months training 29 potential

POTENTIAL LEADER'S TRACK

1. **Church membership** at Victory only requires that you sign a membership application, on which you affirm you are born again. Care Pastors do careful follow-up to new members on the Monday or Tuesday after they sign their application.
2. **Complete volunteer's application** including criminal records check authorization.
3. **Foundations Class** — this seven-week class on Sunday mornings is similar to a new member's class, and it gives the basic house vision, taught by Daugherty over video.
4. **Cell leadership and manual training** — this eight-week, mid-week class or intense Saturday training (which has audio tapes for further instruction) covers both cell leadership principles and the nuts and bolts of cell ministry.
5. **On-the-job training** as an apprentice, usually for six to eight weeks, can be taken simultaneously with the training.

leaders weekly during the Sunday School hour. He later met with his leaders every Tuesday night for 90 minutes, teaching that week's cell lesson and modeling to them the best way to apply the "five-fold vision." When Daugherty started to travel for ministry in various places, he placed Jerry and Lynn Popenhagen over those weekly training meetings.

Since 1991, when Daugherty began traveling monthly to minister overseas, the basic training material has remained the same, while the process has expanded. Potential leaders now must be members of the church, go through an eight-week Foundations Class, and complete an application that authorizes a criminal records check. After they go through Victory's cell leadership training, all leaders meet monthly with their Care Pastors or department heads. Once a month, there is a combined leaders' meeting with Daugherty, usually scheduled before a Wednesday evening service. While some departments, such as Christian Education and Outreach, have their own training, the greater bulk of Victory's cell leaders receive training through the sessions offered by the Pastoral Care Department.

Core Teaching

Since 1983, eight sessions have been central to Victory's core teaching to cell leaders. Each session starts with a joint teaching segment to cover one of the eight core topics. After each session's teaching, potential leaders form small discussion groups where they apply the material, and "tips" are taught and practiced. Those who will lead zip code or workplace cells are placed in discussion groups according to district zip code areas, and those who will lead specific target cells join in specific target discussion groups.

What does Victory's cell leadership training include? The following eight sections give an overview of these powerful core teachings.

The Believer's Ministry

Victory clearly teaches that it is the believer's job — not that of paid professional clergy — to do the work of the ministry. This first session presents the biblical foundation for the believer's ministry, with emphasis on Ephesians 4:11-12, which shows that those with five-fold ministry gifts or offices are to "equip the saints for the work of the ministry." Believers who minister under the authority of the local church have four benefits: 1) maturity; 2) stability; 3) integrity/character; and 4) growth.

Trainees hear the story of Bob and Shirley Morton, who started Victory's first group in 1983. They experienced the anointing of God "almost like a tangible liquid anointing" even during their first meeting, because they flowed under their senior pastor's authority. In that first meeting, they began to operate in the spiritual gifts for the first time in their lives, and they saw God do wonderful things.

The Mortons' story illustrates how God transforms ordinary people into extraordinary disciples, if they will obey Him and team up with the leadership of the local church. Central in this first session is the theme of destiny: like Esther, God has placed each believer in a specific sphere of influence to minister to others "for such a time as this" (Esther 4:14).

KEY TRAINING SCRIPTURES

Following are key Scripture passages used in Victory's core teaching used to train cell leaders.

On the believer's ministry:

Epheisans 4:11-12 And He gave some, apostles; and some, prophets; and some, evangelists; and some, pastors and teachers; *for the perfecting of the saints, for the work of the ministry, for the edifying of the body of Christ.*

2 Corinthians 3:6 Who also hath *made us able ministers* of the New Testament; not of the letter, but of the spirit: for the letter killeth, but the Spirit giveth life.

2 Timothy 1:9 Who hath saved us, and *called us with an holy calling,* not according to our works, but according to his own purpose and grace, which was given us in Christ Jesus before the world began,

Luke 4:18-19 *The Spirit of the Lord is upon me,* because he hath anointed me to preach the gospel to the poor; he hath sent me to heal the broken-hearted, to preach deliverance to the captives, and recovering of sight to the blind, to set at liberty them that are bruised, to preach the acceptable year of the Lord.

Spiritual and Motivational Gifts

Session two shows how God gives gifts to the believer to help him accomplish his God-given task. The leader is to put a demand on the gifts of the Spirit within him, and to help each person in his group identify and operate in the gifts God has given him.

> **ON SPIRITUAL AND MOTIVATIONAL GIFTS:**
>
> **1 Corinthians 12:1, 8-10** Now concerning spiritual gifts, brethren, I would not have you ignorant . . . For to one is given by the Spirit the *word of wisdom*; to another the *word of knowledge* by the same Spirit; to another *faith* by the same Spirit; to another the *gifts of healing* by the same Spirit; to another the *working of miracles*; to another *prophecy*; to another *discerning of spirits*; to another *divers kinds of tongues*; to another the *interpretation of tongues*.
>
> **Romans 12:6-8,** (NIV) If a man's gift is *prophesying*, let him use it in proportion to his faith. If it is *serving*, let him serve; if it is *teaching*, let him teach; if it is *encouraging*, let him encourage; if it is *contributing to the needs of others*, let him give generously; if it is *leadership*, let him govern diligently; if it is showing *mercy*, let him do it cheerfully.

This session gives a concise definition and a clear explanation of each of the nine spiritual gifts in 1 Corinthians 12:8-10, and of the seven motivational gifts in Romans 12:6-8. For example, tongues is "speaking supernaturally in a language not known by the individual, an evidence of the indwelling work of the Holy Spirit." Mention is then made of the three types of "tongues" in the New Testament: 1) the tongues recorded on the day of Pentecost, a known language unknown to the speaker (Acts 2:4-11); 2) the public use of tongues, which must be interpreted to be meaningful (1 Cor. 12:10); and 3) the private use of tongues in the context of one's prayer life, reflected in the words of Paul in 1 Cor. 14:14-15, 18-19.

The Word of God and Prayer

Session three revolves around God's promises for His Word, which is authoritative, reliable, and powerful — an "incorruptible seed" which must be mixed with faith. This session gives nine benefits of reading God's Word, such as the benefits of joy, assurance of salvation, and increased spiritual strength. Included are tips on studying and reading God's Word, as well as four ways to apply what one reads: believe it, speak it, meditate upon it, and act on it. Victory also has an annual Bible reading plan to help each person read through the Bible in a single year.

ON THE WORD OF GOD AND PRAYER:

2 Timothy 3:16 All scripture is given by inspiration of God, and is profitable for doctrine, for reproof, for correction, for instruction in righteousness: that the man of God may be perfect, thoroughly furnished unto all good works.

Hebrews 4:12 For the word of God is quick, and powerful, and sharper than any two edged sword, piercing even to the dividing asunder of soul and spirit, and of the joints and marrow, and is a discerner of the thoughts and intents of the heart.

Luke 18:1 Men ought always to pray, and not to faint.

I Thessalonians 5:17 Pray without ceasing.

Ephesians 6:18 Praying always with all prayer and supplication in the Spirit, and watching thereunto with all perseverance and supplication for all saints.

I John 5:14 And this is the confidence that we have in Him, that, if we ask any thing according to His will, He heareth us.

The segment on prayer stresses the leader's personal prayer life, best when it is daily and consistent. Intimacy with God in prayer helps the leader be sensitive to God's Spirit and His promptings. This is vital; group members need to see the leader as a person of prayerful faith in God's Word. The 'model' prayer in Matthew 6:9-13 is taught, as are practical tips for effective prayer. "Prayer," declares Jerry Popenhagen, "is the fuel that drives the engine of your cell group."

Potential leaders are taught that when a person does not receive answers to persistent prayer, one frequent reason is that he is not praying God's Word. It is therefore crucial that all prayer be rightfully based on God's will, clearly reflected in His Word.

Victory teaches potential leaders that those in a single cell group might be at different levels in their faith, in their understanding of God's Word, and in their prayer lives. But, no matter what the differences, the leader must affirm and value each one, and help each person feel like an important part of that cell.

Spiritual Authority

The fourth session highlights the importance of reigning in life through Christ. The leader can reign in life as he shapes his thoughts and behavior around God's Word, moves beyond past failures, forgives, and gets rid of strife. This can only happen as the leader takes his rightful spiritual authority; at any moment we operate in one of two spiritual kingdoms — the Kingdom of God ruled by Jesus Christ, or the kingdom of darkness ruled by the enemy.

ON SPIRITUAL AUTHORITY:

Ephesians 6:11-12 (NKJV) Put on the whole armor of God, that you may be able to stand against the wiles of the devil. For we wrestle not against flesh and blood, but against principalities, against powers, against the rulers of the darkness of this age, against spiritual hosts of wickedness in the heavenly places.

Romans 5:17 (NKJV) For if by the one man's offense death reigned through the one, much more those who receive the abundance of grace and the gift of righteousness will reign in life through the One, Jesus Christ.

I John 4:4 (NKJV) You are of God, little children, and have overcome them, because He who is in you is greater than he that is in the world.

Colossians 1:13-14 (NKJV) Who hath delivered us from the power of darkness, and hath translated us into the kingdom of His dear Son, in whom we have redemption through His blood, the forgiveness of sins.

Colossians 2:15 (NKJV) Having spoiled principalities and powers, He made a show of them openly, triumphing over them in it.

Romans 8:37 (NKJV) Yet in all these things we are more than conquerors through Him Who loved us.

Spiritual authority is defined as the "power to stand in the name and authority of Jesus and enforce His will (God's Word) over spiritual and physical powers or circumstances." It is God's plan that man be under His authority and exercise dominion in the earth. By his sin, man lost spiritual authority. Jesus restored the believer's spiritual authority through His victory at Calvary over sin, death, and the enemy.

That spiritual authority must be exercised. The believer who walks in spiritual authority has the same power that raised Jesus from the dead. However, if one fails to take his spiritual authority, there is a consequence. "If you do not operate in the spiritual authority God has given you," participants learn, "the enemy will take it away from you and use it against you. In today's wording, 'Use it or lose it.'"

This is one reason behind "prayer walks." As one is walking and praying in a community or workplace, one is taking spiritual authority over territorial spirits. The cell leader is to remind those in his group to keep taking spiritual authority; the enemy does not have the right to dominate any believer's life.

Victory teaches that righteousness, or godly virtue, is necessary for spiritual authority, and is made possible because the believer is "the righteousness of God" in Christ Jesus (2 Cor. 5:21). We exercise spiritual authority through the rightful use of the name of Jesus and the Word of God, our primary weapons against the enemy. The leader also uses spiritual authority when he shares his testimony, claims the power in Jesus' blood, and agrees with another believer in prayer. Another important weapon is the power in praising God, for praise brings God's presence into every situation and circumstance.

Leadership Integrity

If a leader has integrity, his character is solid, complete, and whole. This fifth session points out that the leader's words and deeds must match, no matter whether that leader is at home, at work, at church, or by himself. Integrity pleases God (1 Chron. 29:17), and brings God's protection (Ps. 25:21; Prov. 10:9) and guidance (Prov. 11:3). Scripture also points out that a leader should not be a recent convert (1 Tim. 3:6), because he first requires a time of testing and proving. A leader of integrity can truthfully say, "I am who I am no matter where I am or who I am with."

GODLY CHARACTERISTICS FOR FEMALE LEADERS OF INTEGRITY

Titus 2:3-5 (NIV) . . . reverent in the way they live, not to be slanderers or addicted to much wine, but to teach what is good . . . self-controlled and pure . . . subject to their husbands, so that no one will malign the word of God.

> **GODLY CHARACTERISTICS FOR MALE LEADERS OF INTEGRITY**
>
> **I Timothy 3:2-7** (NIV) . . . must be above reproach, the husband of but one wife, temperate, self-controlled, respectable, hospitable, able to teach, not given to drunkenness, not violent but gentle, not quarrelsome, not a lover of money. He must manage his own family well, having his children in submission and with all reverence . . . He must not be a recent convert . . . He must also have a good reputation with outsiders.
>
> **Titus 1:7-9** (NKJV) . . . must be blameless, as a steward of God, not self-willed, not quick-tempered, not given to wine, not violent, not greedy for money, but hospitable, a lover of what is good, sober-minded, just, holy, self-controlled, holding fast the faithful word as he has been taught, that he may be able, by sound doctrine, both to exhort and convict those who contradict.

The leader of integrity understands that his senior pastor and pastoral staff are fallible, yet he still respects their leadership and stays under their spiritual authority. The leader of integrity must especially guard his speech and not talk against others. Victory teaches potential leaders, "Assist, don't analyze; support, don't critique; serve, don't judge; undergird, don't undermine. Be a Body-builder, not a fault-finder."

A leader with integrity is a team member who is faithful, even in the little things (Luke 16:10). Victory asks each leader to attend at least one service on Sunday and Wednesday, and asks each leader or someone in his group to complete Victory's monthly "cell group attendance sheet" and mail or fax it to the church. Leaders are asked to help with altar ministry by completing an "altar call ministry data card" on a specific person at the altar and praying with that person. Leaders are also to follow-up on these data cards, as well as on the new member cards from those in their area or group type.

Spiritual Warfare

In the sixth session, Victory teaches that the effective leader must engage in spiritual warfare, because demonic forces still try to hinder people today. Jesus came to bring abundant life, while the enemy comes to steal, kill, and destroy (John 10:10). Jesus Himself went through spiritual warfare during His wilderness fast, when the enemy tempted

Him without success (Lk. 4:1-12). As God sent Jesus to bring release and freedom in the lives of those the enemy had captured, so the group leader must help bring breakthrough for others.

The enemy has three basic sources of temptation. One source is the "lust of the flesh," which includes fleshly appetites like sexual sins, gluttony, and substance abuse. Another source, the "lust of the eyes," involves sins like greed, covetousness, and stealing. "Pride of life" is the third source of temptation, with sins like vanity, strife, idolatry, jealousy, and anger that center on the self (1 Jn. 2:16).

There are three strategic "battlefields" each leader must fortify. First is the battlefield of the mind, where the leader must bring every thought "captive" to the obedience of Christ (2 Cor. 10:5). Second is the battlefield of the heart. The leader must guard his heart — including his attitudes and emotions — for his heart is a "wellspring" of life (Prov. 4:23). Third is the battlefield of the mouth. The leader is to make sure his speech is godly, in line with God's Word, and not reflecting doubt and unbelief (Prov. 18:21; Ps. 141:3).

Jesus Himself is our Advocate, sitting at the right hand of His Father, making intercession for the believer (Rom. 8:34). A leader's spiritual weapons include the Word of God, faith, continual prayer, truth, righteousness, the Gospel of peace, and assurance of salvation (Eph. 6:14-18). The leader can increase his spiritual "fitness" as he prays and listens to God, meditates on God's Word, fellowships with godly believers, prays

ON SPIRITUAL WARFARE:

1 Peter 5:8 (NIV) Be self-controlled and alert. Your enemy the devil prowls around like a roaring lion looking for someone to devour.

Ephesians 6:12 (NKJV) For we wrestle not against flesh and blood, but against principalities, against powers, against the rules of darkness of this world . . .

2 Corinthians 10:4-5 (NIV) The weapons we fight with are not the weapons of the world . . . We . . . take captive every thought to make it obedient to Christ . . .

Proverbs 18:21 (NIV) The tongue has the power of life and death, and those who love it will eat its fruit.

Proverbs 4:23 (NIV) Above all else, guard your heart, for it is the wellspring of life.

in the Spirit, and worships God.

To help in this warfare, Victory's goal is to have every leader, every apprentice, and every potential leader go through an "encounter weekend." These encounters are essential, for a leader must confess, repent, and deal with things in his own life before he can minister effectively.

SPIRITUAL WARFARE PRAYER

I will put on the whole armor of God, that I may be able to stand against the wiles, schemes, plots, and wicked plans of the devil. I do not wrestle against flesh and blood, but against principalities, against powers, against the rulers of the darkness of this world, against spiritual wickedness in high places.

I will be sober and vigilant, because the adversary, the devil, walks about as a roaring lion, seeking whom he may devour.

In the battlefield of the mind, I will walk in the flesh but I do not war according to the flesh. The weapons of our warfare are not carnal, but mighty in God for pulling down strongholds, casting down imaginations, and every high thing that exalts itself against the knowledge of God. I will bring my thoughts into captivity to the obedience of Christ.

In the battlefield of the heart, I will keep my heart with all diligence, for out of it springs the issues of life. I will guard my attitudes and emotions.

In the battlefield of the mouth, I will remember that life and death are in the power of the tongue and those who love it will eat its fruit. Set a guard, O Lord, over my mouth and keep watch over the door of my lips. I will not speak doubt and unbelief.

I will not allow Satan to tempt me with the lust of the flesh, the lust of the eyes, or the pride of life. I will put on the whole armor of God, that I may be able to stand in the evil days.

I will gird my loins about with truth, put on the breastplate of righteousness, and shoe my feet with the Gospel of peace. I will lift up the shield of faith, which will quench all the fiery darts of the wicked one. I will put on the helmet of salvation and use the sword of the Spirit, which is the Word of God. I will pray in the Spirit. I will be strong in the Lord and in the power of His might.

Scriptural basis: Eph. 6:11-12; 1 Peter 5:8; 2 Cor. 10:3-5; Prov. 4:23; 18:21; 1 Jn. 2:16; Eph. 6:14-18, 10.

Delegation — Key to Success

No cell leader can fully care for all the people in his group. A single group might consist of people who range from baby Christians to mature believers. In order to survive and thrive, the cell leader must learn how to delegate or "transfer authority or responsibility from one person to another."

A leader is neither to measure "success" by the size of his group, nor by the length of time that group has stayed together, but rather by how many times his group has multiplied and started other groups. Multiplying allows the leader to train others to do the work of the ministry. In order to multiply his group, a leader must first learn to delegate. As the cell leader delegates, more needs are met.

This seventh session focuses on Moses' example of delegation in Exodus 18:13-26. Moses sat through the day while people stood in long lines to have him judge their cases. The "Moses complex" limits work and ministry to the primary leader, and it results in both the leader and the people becoming worn and weary. Jethro, Moses' father-in-law, advised Moses to teach the people corporately and to delegate his responsibilities to others. In this way, people would bear the load with him, Moses would be able to endure, and all the people would go to their place in peace.

JETHRO'S ADVICE TO MOSES (EX. 18:17-23, NIV)

"What you are doing is not good. You and these people who come to you will only wear yourselves out . . .

"But select capable men from all the people — men who fear God, trustworthy men who hate dishonest gain — and appoint them as officials over thousands, hundreds, fifties and tens. Have them serve as judges for the people at all times, but have them bring every difficult case to you; the simple cases they can decide themselves.

"That will make your load lighter, because they will share it with you. If you do this and God so commands, you will be able to stand the strain, and all these people will go home satisfied."

People want to feel needed. When the leader delegates within his cell, it makes his job easier, provides training to others, shows trust in those to whom he delegates, increases commitment, and gives a greater sense of "ownership" and ministry in that cell. This principle is true at Victory, for the church has delegated personal ministry to the cells: 95% of people who come to Victory's counseling center do not go to a cell. Victory has found that people plugged into a cell group often have problems resolved *before* they turn into a crisis.

Delegated organization at Victory includes Daugherty as senior pastor, cell pastors or "Care Pastors," area coordinators (over about five groups), and cell leaders. Within each group there is to be an apprentice (intern), as well as a wide variety of possible roles — praise and worship leader, intercessor (prayer warrior), greeter, "scribe" (with a prayer journal that includes praise reports), record keeper (of attendance, salvations, healings, etc.), children's leader (Kidz Club Leader), refreshments coordinator, follow-up caller, coordinator of prayer walks and outreach, and discussion leaders. Cell leaders are expected to know their limits and to refer those they cannot fully help for counseling through Victory's Personal and Family Life Ministry Department.

An Anointing for Appointing

Through the years, the topic of this eighth session has shifted from "an anointing to serve" to "an anointing for appointing." While there is an anointing that comes upon a believer when he links up with the "house vision" of a local church (explained more in chapter six), the leader also needs an anointing to identify and raise up other leaders, a teaching of special importance after Victory developed G-12 groups.

This "anointing for appointing" is present in the New Testament. Jesus did not ask for volunteers; He rather prayed through the night and then called specific disciples to Himself (Luke 6:12-13). When there was a problem with the Hellenistic widows being overlooked, the apostles chose or appointed seven to wait on tables (Acts 6:1-6).

The impact of this New Testament "appointing" is clear. The apostles Jesus appointed later rocked the world of that time. Stephen, one of the seven chosen to lead the "target cell" of Hellenistic widows, became the first martyr (Acts 7). Philip, another of that seven, brought the Gospel to the region of Samaria and to an Ethiopian eunuch (Acts 8).

How does this apply to cell leaders? This is a step beyond delegation. Victory teaches that the cell leader must pray and seek the Lord as he looks among the people in his group. When the leader senses God's hand on a person, he should challenge that person and boldly speak confidence and direction into his life. When a leader speaks into someone's life under the anointing of the Holy Spirit, a deposit is dropped into that person's heart that later grows into leadership.

Some cell leaders will not do this because they do not want to "put a burden" on others. According to Daugherty, however, the "greatest thing in the world" is to work for Jesus Christ. When the leader uses his "anointing for appointing," he can help a person move to a place of greater fruitfulness.

In Later Years

In later years, Victory developed a printed *Cell Leader Manual and Training Notebook,* complete with fill-in-the-blank notes. This manual contains expanded teaching on Bible foundations for cell ministry, qualifications of a cell group leader, how to start a cell group, how to conduct a cell group meeting, and how to do follow-up.

Victory sometimes refers to the eight core sessions in the first part of the manual as "cell leadership training," and to practical teaching in the manual's second part as "cell manual training." Victory presents this training in several formats, using eight to twelve consecutive Wednesday nights an hour before the midweek service, or other weeknights.

But Victory's most popular training format is its intense Saturday 9:00 A.M. to 2:00 P.M. training, with four hours of teaching divided by an hour for a catered lunch. Care Pastors teach the practical "how to" aspects of cells from the manual in those four hours, and potential leaders take home the audio series of the eight "core teachings" on cell leadership.

On a warm Saturday in July, I attended one of these intense training events. Three Care Pastors taught a wide variety of practical material that covered 80 pages in the manual, sharing inspiring testimonies and their own experiences. Especially highlighted was developing others as leaders and having God's heart for the lost. Other topics included the benefits of a G-12 group, a cell leader's requirements and job description, the importance of prayer and the Word in the leader's life, the five-fold vision, goal-setting, creative teaching tips, and completing the cell attendance sheet and other forms.

During this training I sat among 28 potential leaders gathered around six large tables. Robert and Sharon planned to start a zip code cell after being in Victory for one and a half years. Sandy determined to start her group as soon as her apprenticeship ended. DaRonn and Angie wanted to lead a "Couples' Connection" cell. Earline served as an apprentice in a Sunday cell. Sarah, a 62-year-old grandmother, planned to start a prayer fellowship cell.

The Care Pastors concluded with spirited prayer, laying hands on each one. Some began to weep, a few fell to the floor, but all were touched by a sweet sense of God's presence. This emphasis on prayer and ministry, as well as on teaching, permeates Victory's cell training.

Seeding and Sending

How effective is Victory's training? One example is Jerry and Jan Hauser, raised from the ground up to be Care Pastors. Like all prospective leaders, Jerry and Jan Hauser first became church members, completed the eight-week Foundations Class, and then filled out the church's volunteer application.

The Hausers, like most potential leaders, found the intense Saturday training most practical for their busy schedules. The Hausers' next step was on-the-job training, when they served as apprentices in a group in Tulsa for six weeks. They led their first group in Bartlesville, an hour from Tulsa, and were discouraged by the initial low response. They grew that group by reaching out to a handicapped boy and a woman going through cancer treatment, soon more people joined that growing cell.

After two years, the Hausers turned the group over to their apprentice, moved to Tulsa, and planted another group in their new apartment. In 1991, after Hauser completed Victory Bible Institute (VBI), he and Jan came on staff as Care Pastors. Like the Hausers, most Care Pastors were first effective group leaders, and some even functioned as area coordinators. The Hausers now serve 380 families and 22 cell groups in their district.

PERSONAL PROFILE OF VICTORY'S TYPICAL CELL LEADER

AGE: 36-45

GENDER: Female or a couple*

MARITAL STATUS:
Married, never divorced

FAMILY STATUS:
One child still at home

EDUCATION COMPLETED:
Associate Degree

ANNUAL HOUSEHOLD INCOME:
$40,000-$49,000

SELF-EMPLOYED: No

FIRST TIME TO LEAD A GROUP:
Yes

LENGTH OF TIME A LEADER:
16 months

PERCEIVED GREATEST BENEFIT OF GROUP:
Development of relationships

BELIEVE GOD HAS PLACED OUR SENIOR PASTOR IN OUR CHURCH AS OUR SPIRITUAL AUTHORITY:
Strongly agree

EXPECTATIONS OF THE HOLY SPIRIT IN GROUP:

1) Convict of sin (88.6%)
2) Bring healing and deliverance (87.6%)
3) Bring a sense of Christ's peace and joy (80%)
4) Manifest gifts of prophecy, tongues, words of knowledge... (77.1%)

SOURCE: Representative sample survey taken July 2000

**Source: Pastoral Care Department*

Victory has trained hundreds more leaders than they have kept. Because many of their first group leaders were university students, Victory lost half of their group leaders in the summer of 1984. Most went to different cities and states; the Popenhagens and others were shaken. Then one wise staff pastor gave this view: Victory was not losing them, but 'seeding' them into other churches and areas.

Victory is indeed a sending church, concerned with much more than its own congregation or even its own city. Even today 25 of the first 29 "Bible Fellowship" leaders are in full-time ministry, and dozens of others are pastors of churches scattered throughout America or missionaries in other lands.

CHAPTER FIVE: SMALL GROUP DISCUSSION

Icebreaker: Briefly share about a person (other than a family member) who molded your life in a positive way. What is one leadership quality that person had?

Questions about this chapter:

1. In your own words, describe how Victory prepares a person to be a cell leader.
2. Why do you think Victory's training starts with focus on the "believer's ministry," then covers spiritual and motivational gifts followed by a session on the Word of God and prayer? What priority does prayer have?
3. Consider the link between Victory's sessions on spiritual authority, leadership integrity, and spiritual warfare. What impressed you most in reading about these? Why?
4. From your viewpoint, share why you think Victory takes two sessions to teach about delegating and appointing.
5. Of all the insights Victory has learned about training through the years, which one did you find most interesting? Why?

Application: Share one principle or practice you gleaned from this chapter on Victory's training. How could that be best integrated into your present or future cell group? Your church?

6

Running with the Vision

Victory's Pastoral Staff and Its Role in the Cell Groups

The vision Daugherty has instilled in Victory is to plant a cell group in each of Tulsa's apartment complexes, neighborhoods, schools, and workplaces — meeting wherever and whenever possible. It is a vision captured in one of Victory's many mottos: "Building God's Kingdom, one cell at a time."

This is a deeply personal vision. Each member of Daugherty's family leads or helps out with at least one cell group: Daugherty heads a leadership G-12 group; wife Sharon leads weekly music choir cells, a monthly group for staff pastors' wives, a traveling ministers' cell, and the weekly Thursday morning Bible study; oldest daughter Sarah is the Junior High Pastor and leads a junior high co-ed cell; Ruthie leads a senior high cell, band cell, and youth cell, and is assisted by brothers John and Paul as Children's Pastor of the 10:00 A.M. service.

"There's a time when the vision comes," Daugherty declares, "you need to run with it. Habakkuk 2:2 states, 'Write the vision, and make it plain upon tables, that he may run that readeth it.'"

House Vision

Talk with the Popenhagens, and it won't be long before they mention the term "house vision." They repeatedly stress that staff members are not to have a separate agenda but are to stay in line with the vision God has given their senior pastor.

This attitude flows through all levels of leadership. With the kind help of the Pastoral Care Department, I collected survey results from a representative sampling of 105 Victory cell leaders. More than nine out of ten strongly agreed with the statement: "I believe God has placed our senior pastor in our church to serve as our spiritual authority, and his vision is to be carried out in decisions regarding our cells' direction and materials."

This view breeds unity. Victory's staff quickly agrees with Daugherty that leadership requires a "Gethsemane," dying to one's own desires, in order to fulfill God's plan and designs. But this attitude does not end there.

"We believe that God has a plan and a purpose for each believer," states Popenhagen. "According to Ephesians 2:10, we are 'created in Christ Jesus unto good works, which God hath before ordained that we

H onor the man of God to whom God gave the house vision.
O bey the directives given for the cell ministry.
U nderstand the heart of the ministry.
S ubmit to the calling of God to administrate this vision.
E njoy what you are doing and be enthusiastic.

V alue the call of God on each believer's life.
I nvest your time to equip believers for the work of the ministry.
S erve as unto the Lord.
I mplement the cell ministry in accordance with the pastor's vision.
O vercome obstacles that arise in the cell ministry.
N ever be negative or pessimistic in regard to the senior pastor or any part of the house vision.

Written by Jerry and Lynn Popenhagen

should walk in them.' Much of our job is to help birth that in each believer's life, and help believers discover the works that God has already prepared for them to do. There is such a joy and a purpose in equipping people for ministry."

CARE PASTORS

I am a curious person, often about things that other people are too busy to consider. I was impressed by Daugherty's vision, the diversity of Victory's cells, how training was done, and the attitudes expressed. But what I wanted to understand most was Victory's pastoral staff. Pastoral staff are the vital "link" who determine whether or not vision will become reality, who do the actual training, and who carry out the daily grind that makes or breaks the "cell church dream."

At the time of this writing Victory has 31 full-time pastoral staff members, each of whom oversees or helps with the cell groups in his department or area of ministry. I personally interviewed one staff pastor in each of seven different departments, looking for one "type" to focus on and explore. I found that no staff pastor better typifies Victory's heart for cells than the Pastoral Care Department's six Care Pastor couples, who fervently "run with" the house vision.

Each Care Pastor is over a specific geographical Tulsa district, including that district's cell groups, as well as the member families who reside in that district, whether part of a cell or not. So what are these Care Pastors like? Above all, I found them well experienced in cell life, in ministry, and in cell priorities.

Consider the Shouses. During the two years Howard and Cheryl Shouse served as area coordinators, they multiplied their group 19 times. In that time the Shouses saw five marriages restored, nine people without transportation given cars, multiple salvations and healings, and 100 people find jobs. Even though Howard remains an accountant/auditor with Oklahoma's Tax Commission, he joins Cheryl as a Care Pastor serving 359 families and 55 cells involving 700 people. Many zip code cell leaders still call the Shouses "my spiritual Dad and Mom."

Al and Maria Leerdam, originally from the Virgin Islands, are loved by the 334 families and 23 cells in the district they serve. Leerdam is quick to point out the need for training. "The harvest is ripe," Leerdam emphasizes. "We can ask God to send the laborers out, but we must first help prepare them."

Don and Arlene Hanson, the newest of the Care Pastors, serve 426 families in their district, and oversee 20 zip code, business, and target groups. Like most Care Pastors, this attractive African-American couple once led a group and has seen prayers answered. The Hansons prayed for one member in their group of disabled people who had a hole in her heart; by the time she was discharged from the hospital, she had an x-ray proving her heart was healed and whole. Another time, their group prayed for Wendy, who had cancer, a bald head, and suicidal thoughts. Within a few months Wendy had a full head of hair, felt healthy and strong, and was happy to be alive.

Contact, Connect, Train, Resource, and Minister

What do pastoral staff do? I spent the better part of a week interviewing and observing Victory's pastoral staff, with a continued focus on Care Pastors. Each seemed to have five basic aims: contact, connect, train, resource, and minister.

Contact

Each service Care Pastors stand and greet people at the doors. They are warm and friendly, and they pay special attention to people they do not know. Whether they walk through a worship service crowd, or among people gathered for events like the monthly New Members' Dinner, Care Pastors and Victory pastoral staff are skilled at meeting new people and making caring contacts. I watched one Care Pastor personally meet 34 people at a New Members' Dinner, careful to repeat each name shared.

Connect

But Care Pastors do not consider it enough to make a caring contact; they connect the uninvolved into groups, and the interested into leadership training. I stood near one Care Pastor on a Sunday morning. Within 20 minutes he had told two visitors about Victory's membership application and upcoming New Members' Dinner, thanked four cell leaders for their hard work, greeted five cell members, invited three interested members to that Saturday's cell leadership training, and challenged a former leader to start another cell. Care Pastors also "connect" people every Monday when they make follow-up phone calls and visits to visitors and altar respondents from the day before.

Train

Care Pastors rotate and teach portions of Victory's cell leadership and manual training. Each Care Pastor has at least one weekly or bi-weekly leadership G-12, where he does the bulk of his ongoing training. Care Pastors also observe cell groups and train by sharing insights with leaders. After one cell meeting, I listened as a Care Pastor talked to the leader couple, discussing options for dealing with a difficult person in their group. They chose the option he felt best, and soon after, their cell began to grow.

Resource

Victory gives key resources to Care Pastors, including subscriptions to *CellGroup Journal*, a quarterly magazine designed for group leaders. Care Pastors share the best and most relevant of these resources with leaders. When staff members spot good books on cells, Victory purchases them in large quantities as gifts for their leaders. Care Pastors also resource leaders with specific needs. For example, I heard one Care Pastor refer a troubled person, who needed more help than the leader and group could give, to Victory's Personal and Family Life Department for counseling. Another Care Pastor gave one of his books on teaching to a frustrated leader.

BEST BOOKS USED AS RESOURCES AT VICTORY

Successful Home Cell Groups
by David Yonggi Cho

Where Do We Go From Here?
by Ralph Neighbour, Jr.

Home Cell Group Explosion
by Joel Comiskey

Groups of 12
by Joel Comiskey

Source: Jerry Popenhagen

Minister

The supreme role of Victory's pastoral staff, including each Care Pastor, is to minister. Care Pastors minister to families in their district who are not yet in a cell and encourage them to get involved as soon as possible. Care Pastors minister to altar respondents near the end of every worship service. They pray and minister to leaders in their G-12 meetings, during telephone calls, home visits, and their many weekly contacts. These models of ministry inspire by their example, and have helped instill a spirit of ministry in Victory's cell groups. One Care Pastor seemed to have a special ministry of encouragement. Her every comment built confidence and pointed to God's Word. The cell groups she oversaw were the same, quick to pray for needs and speak words of encouragement.

AREA COORDINATORS

But no Care Pastor can do his work alone; groups need a caring structure to thrive and grow. After Care Pastors were hired in 1990, they placed seasoned volunteer leaders as "area coordinators," over up to five cell leaders or groups. The area coordinator's job was to care for leaders, help solve problems, and assist those groups in evangelism and multiplication. Some area coordinators also led a cell; others did not.

Todd and Charly Young are well qualified to serve as area coordinators. In the 12 months they were group leaders, five of their members started other groups. The Youngs, who now serve as area coordinators over four cell groups, regularly call, pray, and meet with their cell leaders. They assist with any church sponsored activity in their area, and help group leaders evangelize door to door or plan upcoming events. "If I were to give advice to a new area coordinator," Young stressed, "it would be this: 'Pray continuously, reach out, and let your leaders and people know you are there to help them. Read the Word and attend church services two to three times a week to keep getting refilled so you don't get empty by giving out. Be supportive to the leaders above you, especially to Pastor Billy Joe's vision and leadership. And never quit or give up.'"

G-12 AND TRANSITION

In 1998, Victory began the transition to leadership G-12 groups, a group of leaders who, in turn, lead their own cell groups. Instead of forcing Care Pastors to automatically have their area coordinators in their G-12 groups, Victory allowed each Care Pastor to choose who he wanted in his G-12. The results? Choice allowed more unity in a G-12 group, but it also meant that most districts had both a leadership G-12 and a few area coordinators.

"Transition's tough," one Care Pastor told me. "We had wanted to go entirely to G-12 groups. We thought they could be more effective. But not long after we took some of our leaders to G-12 groups, we found that our number of zip code cell groups began to drop. While G-12 has sharpened our focus on mentoring and raising up new leaders, we realized that G-12 was not growing our districts, so we have slowed the transition. At this point many districts still have some area coordinators from the former system, as well as G-12 leadership groups."

CHAPTER SIX: SMALL GROUP DISCUSSION

Icebreaker: Of all the staff pastors you know, who do you most admire? Why?

Questions about this chapter:

1. How would you explain "house vision" to someone who has never read this book nor been to Victory?
2. Why do you think the author considers staff pastors so important?
3. What basic five things do the pastoral staff do?
4. Consider a system with area coordinators, and one with leadership G-12 groups. Which one do you prefer? Why?

Application: Share one principle or practice you gleaned from this chapter on Victory's pastoral staff. How could that be best integrated into your present or future cell group? Your church?

7

A Day in the Life of a Care Pastor

A Closer Look at What Care Pastors Do and Value

Staff pastors in traditional churches spend most of their days resolving crisis situations, planning future events and programs, doing paperwork, counseling in the office, and taking uninvited phone calls. Staff pastors in the cell church are different. They focus more on raising up, equipping, and encouraging their leaders, and they are much more "hands on" in their approach.

To learn about the actual day-to-day job of the Care Pastors, I asked if I could "shadow" one for a full day. I knew this would provide valuable insight into what Victory's Care Pastors do and value.

I was paired with Eric and Melody Castrellon. The Castrellons had thought they would live and die as youth pastors in rural Arkansas. But when they visited Tulsa and went to see Victory's monthly crusade in a low-income area, they were impressed with "Pastors Billy Joe's and Sharon's hearts for the lost." When they came on staff in 1995 to pastor Victory's young singles, Eric and Melody were delighted, and were soon busy raising up singles' cells and cell groups on college and university campuses.

On this July Monday in 2000, Eric and Melody are marking their eighth month as Care Pastors in the Pastoral Care Department. The

Castrellons serve 460 Victory families in west Tulsa, and are over 45 cell groups: 14 married couples' cells, 15 zip code cell groups, 12 business and workplace cells, and four young marrieds' home-based cells. Their lives, they say, are full of ministry. One thing I was to learn: Eric and Melody understand their priorities. They don't treat groups as a machine to be managed but as a garden to be tended.

TYPICAL WEEKLY SCHEDULE OF A CARE PASTOR
Minimum: 45-50 hours

SUNDAY:
Greets at assigned door and helps with altar ministry in four services, at 9:00, 11:00 A.M. and 6:00 P.M. in the Mabee Center, and at 10:00 A.M. in the west campus. Shakes the hands of at least ten new people or couples, and makes a minimum of seven meaningful contacts — to discover how he can better help and resource those persons, and better connect those persons to cell groups or ministry involvement.

MONDAY:
Takes follow-up referrals left in Mail Center of visitors, new members, new converts, and those needing prayer, and begins follow-up as soon as possible.

TUESDAY:
Meets with the Popenhagens and other Care Pastors from 9:00 to 11:00 A.M., and then joins in staff prayer with Daugherty from 12:30 to 3:00 P.M. Tuesday evening is taken with visits to cell groups or home visits.

WEDNESDAY:
Makes daytime pastoral and cell visits. Greets and helps with altar ministry in the 7:00 P.M. service.

THURSDAY:
Time in office and miscellaneous activities and projects.

FRIDAY:
Day off to spend relaxing with family.

SATURDAY:
Makes pastoral and cell visits.

Monday through Thursday:
Each Care Pastor leads one G-12 meeting a week, observes at least one group meeting, and keeps in contact with area coordinators. Care Pastors are also responsible to make hospital visits, as well as to officiate and help with weddings and funerals.

Prayer Journey — 8:45 A.M.

Eric's day begins as he meets the Popenhagens and other Care Pastors in multi-purpose room 254 of the east campus building. Melody is at home, caring for their two young daughters.

Victory has a metroplex-wide vision for the city of Tulsa. Care Pastors not only minister, they also serve as 'spiritual gatekeepers' of their areas. When the Popenhagens discovered a decline of 33 geographical groups over 18 months, they knew they had to do something. Inspired by Ted Haggard's example in Colorado Springs, they had staff take 'prayer journeys' in their geographical districts. Today will be their third such journey.

God gave each tribe of Israel a different area to take, but they still fought together to take each area. Based on this example, Care Pastors jointly meet together on special Monday mornings to accompany one Care Pastor on a prayer journey in his district. "We have already seen positive results," comments Jerry Popenhagen, "with many former leaders again asking to lead a group, and new business groups started, even in this summer month."

This morning we will go on a "prayer journey" to District 6 in south Tulsa, where Edwin and Delia Miranda serve as Care Pastors over 336 families and 28 G-12, business, and zip code groups. The Mirandas reflect the trend pervasive throughout Victory — ministry is not the domain of the masculine alone; couples are called to minister together.

This prayer journey is tailored to its setting. In Oklahoma's land rush days, people claimed their land by driving stakes in the ground. Before we go, Edwin Miranda displays one of his wooden stakes, inscribed with Joshua 1:3 and 2 Chronicles 7:14-15, symbolic of a spiritual declaration to be placed in strategic points of his district. The staff plans to anoint key intersection points with oil, and to pray that God will raise up cell groups in apartment complexes. After a time of prayer, we are ready to leave.

9:15 A.M.

Our prayer journey begins as 16 of us climb aboard a small bus affectionately dubbed our "prayer mobile." Since Eric has the right kind of license and experience, he serves as our driver. On our bus are: Margaret Hawthorne, director of Victory's prayer ministry; Brett Reed, spiritual mapping coordinator; most of the Care Pastors; the Popenhagens; three observers; and myself.

Over the next three and a half hours we make nine stops as we drive and pray around the parameters of Miranda's southern district. We stop at CityPlex Towers where we go to the empty ballroom on 60th floor with a clear panoramic view of the city. Our prayer there is long and spirited, and we join in a verbal declaration over Miranda's district. Eric stops at three strategic points for Miranda to drive in a stake and anoint that spot with oil as we pray. We stop at Jean Reehling's home, host site of a zip code group for three years, and pray a fervent prayer of blessing over her.

We stop at two businesses with groups: Mark and Donna Atwell's Blessed Auto Sales* and Maurice and Sandy's H-V Manufacturing. Over each, we pray that God would bless their groups and greatly prosper their businesses. Eric stops at the marker of the "world's greatest oil pool," opened November 22, 1905, which triggered the growth of Tulsa's oil industry. We pray that God would abundantly pour out the oil of His Holy Spirit over that area and throughout Tulsa.

Throughout our drive Miranda has Eric make brief stops at major intersections. As Miranda pours a small portion of oil onto the pavement, we join him in prayer. Cheryl Shouse comments, "Everything is birthed by prayer. These prayer journeys keep us on the cutting edge."

---

*Note: Mark Atwell's used car lot was selling an average 20 cars a month. On the Tuesday after we prayed for his business, Atwell sold three cars that morning and three more cars that afternoon.

We stop at a mobile home park in the Jenks area and pray blessings over Chuck and Joyce Werner, their three daughters, and their group. By this time, all are hungry and ready for lunch.

Lunch — 12:45 P.M.

Melody and others join us for lunch at a restaurant, where food and fellowship are our next focus. I sit next to Melody and across from Lynn Popenhagen, who says of Eric: "We were glad when Eric joined us on staff. He was faithful and proven, with Victory's vision for cells. Like other Care Pastors, he had gone through Bible college or seminary. He had a love for people, and was a strong exhorter. Eric also has a flair for evangelism. You cannot join this staff unless you are first a soul-winner."

Monday afternoon follow-up — 2:15 P.M.

Eric goes to his office in the Pastoral Care Department to check his e-mail, as Melody checks hers in an adjacent office, holding their five-month-old daughter Annabelle. Eric then walks to the nearby 'Mail Center' and takes prepared names and contact information for Sunday's salvations, visitors, new members, and altar respondents that live in his geographical district. As Eric looks through his pile, he comments, "In addition to salvations, I try to follow-up on three types of people myself: unchurched visitors, new members, and people who need prayer. I give the others to my cell leaders or area coordinators to contact."

2:45 P.M.

Charles, a real estate appraiser, calls. Since he lives in Eric's district, his call is routed to Eric's phone. Charles has gone to Victory for five years, and calls to ask what the cell ministry is about. After Eric finds out that Charles has already gone through Foundations, the new members' class, he encourages him to get involved. Eric invites Charles and his wife to an upcoming cookout in his area and then to the Saturday cell leaders' training. Eric tells him about the volunteer's application and other requirements. Eric asks about his family, then inquires, "Which Sunday service do you usually go to?"

After they make plans to meet next Sunday, Eric ends the call by praying for him, "Lord, let Your grace and discernment rest mightily on Charles and his family." Eric hangs up the phone, then comments, "Charles has great potential. Unless we get Charles and his family involved, God's destiny for them will never be completely fulfilled."

2:55 P.M.

John, leader of a business group, calls about a monthly outreach crusade he is helping coordinate in Eric's district. This crusade will be the combined effort of ten of Eric's cells, complete with a food give-away, a carnival, an evening concert, and a Kidz Club mobile 'sidewalk Sunday School' in the Sand Springs area.

3:00 P.M.

During the next few minutes, Eric compiles his 'to do list': write a character reference letter for someone returning on staff; coordinate a sandwich/hamburger cookout for former cell leaders in his district; visit a new member; drop by a young single adult cell; call specific leaders to check on them.

3:25 P.M.

Eric checks his e-mail for those now in the hospital, and discovers one man from his district. When he returns to the Mail Center he finds three more visitors from his area who need follow-up, and he discovers his copy of *CellGroup Journal* quarterly magazine. Eric then goes to the Resource Center and gets cell lesson books and other resources to give to the leaders he plans to visit soon.

4:30 P.M.

Five-month-old Annabelle starts to cry. Melody had left to go on an errand, briefly leaving a sleeping Annabelle. Eric, the only Care Pastor with young children, quickly goes to see how she is and cradles her in his arms. Lynn Popenhagen's earlier words ring in my mind: "We combine family and ministry. In the ministry you must pace yourself and take time for your family. That's why we have an open door policy both with our staff and our children."

4:40 P.M.

Eric calls a new group leader to find out how he's doing. While Eric speaks words of encouragement, recently returned Melody and I talk. "It can be frustrating," Melody relates, "to balance ministry to the families in our district with ministry to my own family. But there is also much joy as well. Last night as I watched Eric minister with such compassion to people who came down to the altar, I found myself falling in love with him all over again."

Ministry at home — 5:10 P.M.

Eric arrives home, bringing their two-year-old daughter Jubilee with him, and listens to his answering machine. There are several calls from cell leaders that he must return. Melody tends to their two daughters and prepares for an upcoming music practice.

5:15 P.M.

Eric calls to check on a cell leader who recently tried to transition a Kidz Club to an adult zip code group. No one is at home, so he leaves a message. His primary concern: how are they staying 'connected'? Are they receiving the support and care they need from a G-12 group, or do they have the oversight of an area coordinator?

5:25 P.M.
A concerned zip code cell leader calls Eric to discuss the recent and unexpected marriage of a cell member.

5:30 P.M.
Three others join Melody in their living room to practice two lively quartet songs, to be sung during the upcoming "Word Explosion," Victory's largest annual conference. For the next few minutes, we all relax, listen, and enjoy.

6:30 P.M.
The transitioning cell leader returns Eric's call. Things are not going well. During the lengthy call, Eric encourages and commends the leader. They also talk about the leader's family and marriage. Eric suggests that he and his wife join the "Couples Connection" and receive needed ministry. The call ends with spirited prayer.

Hospital visit — 7:30 P.M.
Eric, Melody, and I go to Tulsa Regional Medical Hospital to discover that the man we were going to visit has already been discharged. Before we leave, Eric looks at his list, and discovers the name of a second man from his district. We are soon praying with the wife of Kenneth Brown, a missionary to Peru, who had suffered a cerebral hemorrhage.

Visit to a cell meeting — 8:35 P.M.
We drop in Jeff Mayhill's young single adult cell that meets in an apartment. Seven are gathered, sharing answers to prayer. One woman had received needed finances from an unexpected job. Another reports that a severe family situation was resolved. The leader's grandmother had been confused and was having trouble communicating before they prayed; he reports that the next day she was clear and lucid, and she

continues to be better than before. One man received $10,000 he needed to finance a crusade in an indigent section of India.

Supper talk — 9:20 P.M.

We drive to a restaurant to eat a late supper and talk about ministry. "I love to hear people repeat the sinner's prayer," Eric reflects. "The most important thing is souls being saved." Melody adds, "Cells have revitalized our ministry. It is the most comprehensive way to minister to people I have ever seen. Cell ministry can take a common person and thrust him into his purpose."

"That's right," Eric interjects. "My passion is to see people discipled to the point that they can disciple others. To reach, train, and release, to see people grow to the point they reproduce themselves in others." Eric pauses, then continues, "It's all about people. It's all about relationships."

Our server Jeremy comes to the table. Eric invites him to church, and talks with him about his spiritual life.

The day ends — 10:30 P.M.

Melody and Eric return home. Their day of ministry has ended, and another will soon begin.

CHAPTER SEVEN: SMALL GROUP DISCUSSION

Icebreaker: What is one thing a staff pastor has done for or with you that has been meaningful to you? Why was that important?

Questions about this chapter:

1. Before going into the details of that day, the author writes that Eric and Melody "don't treat groups as a machine to be managed, but as a garden to be tended." Why do you think she would have considered it important to say this?
2. How would you compare the activities of Victory's Care Pastors to the regular activities of staff pastors you have known?
3. What impressed you most about Eric's activities in this chapter? Why?

Application: Share one principle or practice you gleaned from this chapter on Victory's Care Pastors. How could that be best integrated into your present or future cell group? Your church?

8

FLOW LIKE THE MISSISSIPPI

Insights from Victory's Cell System

When I first went to Victory, I was only mildly impressed. They seemed scattered, meeting weekly on four different campuses spread over a four mile radius, with three of their four Sunday services in a rented indoor sports arena. Some of their facilities were older and needed updating.

But when I returned, I discovered that Victory focuses its finances on people and developing leaders. They construct or purchase multipurpose buildings on a cash-only basis, with one building donated by another ministry. Staff explain that a person might not come into any of Victory's buildings, but is likely to walk through the "open door" of a cell that draws his interest or meets a need. Victory's staff now oversees 35 varieties of nearly 1,000 cell groups, with training that has launched more than 4,000 group leaders. Since 1983 Victory has produced weekly cell lessons and monthly cell group directories, with one of the longest continuous "group histories" of any church in America. Through years of persistence, Victory has built nearly 1,000 "open doors" to draw the lost home and the believer into closer fellowship with God and others.

As I watched and asked questions, my reluctance turned into admiration. I came to see Victory as more than just a cell church to be studied. Victory is a love story to be valued. It's the story of the love of Billy Joe and Sharon Daugherty for God, for the lost, and for their congregation. It's the account of a pastoral staff that is passionate about their senior pastors and the people they were called to serve. It's about cell leaders sold out to God, who serve their members sacrificially. It's the chronicle of a congregation who loves every type person in their city of Tulsa — from inner city youth, to nursing home residents, to proper middle class families in the suburbs.

Am I saying that Victory is the perfect cell church, with a cell system fully polished and problem free? Hardly. Victory has its share of problems and frustrations. Because of cutbacks from former days, Victory no longer has a computerized tracking system for its groups; record keeping is done manually. Many Victory members are still not involved in cells, while others are involved in two or more. Even Jerry Popenhagen thinks that Victory's cells need "more prayer and more evangelism, with follow-up not happening to the degree it should." Popenhagen paused, then added, "But understand this: We move with God's Spirit and can't be put in a box. We flow like the Mississippi River. If we hit an obstacle, we go around it."

If you've ever been involved in cells, you've probably hit an obstacle or two yourself. There is much you and I can learn from Victory's success and mistakes to help us "flow like the Mississippi." Consider adding the following twelve insights to your own growing list.

Insight #1
Give Preeminence to God's Word and the Biblical Vision for Cells

What is preeminent at Victory and her cells? Look in any Victory Sunday bulletin, and you will find that week's cell lesson, complete with a primary biblical truth and multiple scripture references. Listen to Victory's cell leaders' training, and you will find scripture most important; throughout the *Cell Leader Manual and Training Notebook* are biblical passages and truths with pages of scriptural confessions for the new leader to declare.

Talk to any staff pastor or cell leader. Each quickly lets you know that he doesn't have all the answers, but God does; His Word is the light to guide the believer's life. This continual focus on God's Word has triggered cell members to report, "I finally feel like I understand what the Bible says."

CELL GROUP TESTIMONY

Shirley Morton was concerned for a young lady in their group who had quit high school, struggled to have clear thinking, and lived on welfare with her mother. When Shirley encouraged her to memorize Bible verses, she cringed. But the young woman decided to try. So she came early to group meetings and would privately recite to Shirley any Bible verse she had been able to memorize.

That young lady began to improve to the point where she could memorize entire chapters of the Bible. She and her mother both went for vocational training. The young lady later got her GED, and today is married to a man in full-time ministry.

Source: Arlene Hanson

Even Victory's name is based on the account in Exodus 17:9-15 when Israel fought the Amalekites. As long as Moses raised his hands holding God's rod, the Israelites won. When Moses became weary, Aaron and Hur stood on either side of him to hold his arms up, a scene that has become a symbol for partnership in prayer. After they won the battle, Moses declared God "Jehovah-Nissi," or God of victory. Victory Christian Center has needed partnership in prayer to win and overcome its many battles.

Victory has already won the battle for a biblical vision for cells. Daugherty and Victory's leaders are quick to point out that Jesus himself discipled a group and that New Testament believers met in the temple and "house to house." Interestingly enough, throughout my twelve days of on-site observation and interviews, I heard this one statement more than any other: "God has called every believer into ministry."

During Victory's national cell conference in 2000, I listened to Daugherty's humorous account of Moses' attitude in Exodus 18:15-16.

When Jethro, Moses' father-in-law, saw how Moses sat from morning to evening to judge the people's cases himself, he asked Moses why he did this. Daugherty, with a deep and self-assured tone, repeated Moses' own words: "'Because the people come unto ME to inquire of God. When they have a matter, they come unto ME; and I judge between one and another, and I do make them know the statutes of God, and His laws.'"

Daugherty then stated that pastors often make the same mistake: "Like Moses, we must learn to share ministry with our people. Otherwise, we will wear out, and the people will never be satisfied. God clearly shows us in Ephesians 4:11-12 that the purpose of full-time pastors is to equip believers for ministry."

Sharon Daugherty summed this same view in her own workshop, declaring that God has called every believer to do the work of the ministry. Joel 2:28 states that God pours His Spirit out on all flesh, not just a few. "God has called each of us to a lifestyle of exhortation and

BELIEVER'S MINISTRY CONFESSION

I am being equipped, prepared, trained, perfected, and made fully qualified for the work of the ministry. As I do the work of the ministry, I will become mature and not tossed to and fro and carried about with every wind of doctrine, and I will not be deceived. I will speak the truth in love, and I will grow up in my Christian life.

God's Word tells me my sufficiency is from God who made me a minister of the new covenant and a minister of reconciliation. I have been saved for a holy calling. I am God's workmanship, created for good works that I will walk in. I will do the works of Jesus, I will preach the Gospel to the poor, heal the broken-hearted, proclaim liberty to the captives, and recovery of sight to the blind, set at liberty those who are captive, and proclaim the acceptable year of the Lord.

I will obey the Great Commission and preach and teach the Gospel, baptize, and make disciples. Today I make a decision to live a holy life. I dedicate every area of my life to Him. My life will become one of maturity, stability, integrity, and spiritual growth. Christ is the head, and I am the body. I am submitted to His control in my life from this day forth.

Note: For biblical basis, refer to Eph. 4:11-16; 2 Cor. 5:18; 2 Tim. 1:9; Eph. 2:10; Lk. 4:18-19; Mk. 16:15.

Source: *Victory's Cell Leader Manual and Training Notebook*, p.16.

evangelism," Sharon affirms, "the work of the ministry is to go on outside church walls."

INSIGHT #2
PREPARE GROUP LEADERS WITH GOOD TRAINING AND HELPFUL GUIDELINES

Ruthie, Billy Joe and Sharon's second daughter, has been in groups much of her life. When she was asked how to start a cell group, this 'next generation' female explained: "Sometimes you have to aggressively recruit people to come to your cell. But often there are people around you so hungry for God and something more in their lives, that all you have to do is simply invite them. If God has put a particular cell on your heart, just step out and do it."

How does Victory help potential leaders "just step out and do it"? After his first failed attempt at groups, Daugherty learned that it was vitally important to prepare leaders with good training and helpful guidelines. As a full-time cell church consultant and speaker since 1985, I have seen dozens of styles of training material. Victory's eight core sessions, with teaching followed by small group discussion and applied practice, is among the best. My

CELL GROUP TESTIMONY FROM A CHURCH IN VICTORY FELLOWSHIP OF MINISTRIES (VFM)

A VFM pastor spotted a heart for Jesus in a 19-year-old girl and encouraged her to start a group. She went directly to her college president and asked for a room to hold a Bible study group. The president responded by giving her the entire second floor of that building, personally attending the group himself, and providing a buffet lunch for all participants.

Not long after she started the group, that girl's pastor received a lengthy anonymous letter. Jerry Peterson, VFM's Director, recalls a synopsis of that letter: "We want to tell you about a little girl in our college we thought was a Jesus freak. We're drug dealers and we used to make fun of her. We go to this college, and went to her Bible study. There we got saved, filled with the Holy Spirit, have walked away from the drug cartel. We are now preparing for the ministry. We can't tell you who we are, because it has not been long enough, and we might be killed if we did. But we want you to know this: your ministry has reached us."

Source: Jerry Peterson, VFM Director

two favorite were the teachings on spiritual authority and "an anointing for appointing." These sessions keep a strong focus on spiritual concepts with a good balance on practical dynamics in the small groups that followed. When paired with "hands-on" time as an apprentice, and ongoing training in a G-12 group, the new leader is well equipped.

From 1983 on, Daugherty was clear about guidelines: a leader is not to give a lengthy teaching but to share a nugget from that week's lesson that leads to interaction, discussion, and mutual ministry. Daugherty explains: "People in a group want to share. They want to be heard."

Between meetings, the leader is to pray, follow-up, involve, and encourage his group members. Before she was a Care Pastor, Arlene Hanson had a ladies' cell group in which most members were shy and hesitant about taking any type leadership. Arlene told her group that God could do the same thing with each of them that He did through her. One lady, who had been afraid of public speaking, later came to Arlene and announced, "Guess what? I set up my apartment so that I had seats in front of me and pretended I was having cell group." Not long after, that lady started her own cell.

FOLLOW-UP HABITS OF EFFECTIVE CELL LEADERS

Telephone Calls: A telephone call gives a personal touch. Make most calls during the evening between 6:00 and 9:00, spreading out the number of calls made throughout the week. The leader should get others in the cell group to help with this.

Notes: Write brief notes with a scripture of encouragement.

Personal visits: These 15-20 minute visits are going the "extra mile" to know others personally and to pray for any needs.

Source: *Victory's Cell Leader Manual and Training Workbook*, pp. 106-7.

INSIGHT #3
THE SENIOR PASTOR MUST PROMOTE CELL GROUPS AND THEIR BENEFITS

With the senior pastor's backing and promotion, cells thrive; without it, they wither and die. Much of Victory's cell success is due to Daugherty's strong guidance, wisdom, and promotion of the cell groups. At almost every service, Daugherty promotes the groups from the platform. His favorite approach is to have a cell leader or member give a testimony, pointing once more to the power of God working through the vehicle of a cell.

Daugherty and his staff have further shaped cell promotion. No member could overlook the cell lesson, based on one of Daugherty's past sermons, in each week's Sunday bulletin or the monthly "cell group directory." Depending on service location, there is always a "cell group booth" either in the foyer or to the side of the auditorium, complete with folders, information, and a place for leaders to turn in one-page written reports — a booth usually manned by a smiling Care Pastor. Care Pastors stand at each entry point, welcoming people to the service, while inside, cell leaders sit in various spots, careful to greet those around them and invite them to their cell group.

Ask any staff member or cell leader, and they are quick to point to the benefits of being in a cell group. One benefit to being in a cell group is a sense of belonging, especially important when so many are lonely, living in neighborhoods where they do not even know their neighbors' names. Another benefit is stability, for when the storms of life come, one has a group of believers who will pray with him, encourage him, and be there for him when he needs them. "Always highlight the benefits of cell groups," Jerry Popenhagen stresses. "People will invest their time in groups if they feel their time is worth it and see groups impact others' lives."

Insight #4
Invest in the Right Kind of Staff

Many years ago I spoke at a church that had made a recent transition to cells. The senior pastor had done such a great job promoting cells that his church's cell system had became the largest in his country. But all was not well. In his zeal to transition to cells, the senior pastor had hastily chosen a staff of "district pastors," many of whom had never led a cell group, nor understood cell life, nor how to give ongoing support and ministry to group leaders. I watched one district pastor corner a man at a drinking fountain and add him to his list of leaders. When I later questioned that staff pastor, he told me he was just trying to meet his job "quota." Four years after that system hit its peak, there were just a handful of groups left. That senior pastor had invested in the wrong kind of staff, more interested in salary than in the godly support of their leaders.

Daugherty has been wise in shaping Victory's pastoral staff. He started slowly, first with the proven characters of Jerry and Lynn Popenhagen. He then gradually invested money, time, and resources into a growing staff who would recruit, develop, and care for leaders and the people they serve. If you glean no other insight from Victory, learn this — the right kind of pastoral staff can build a dynamic cell system that makes hell tremble. On the reverse, staff wrongly done can demoralize and demolish. If your church is smaller, the same principle applies to part-time or volunteer staff.

How do you know you have the "right kind" of staff? Most important is an attitude of unity. Jerry Peterson has been on Victory's staff since 1984. He directs Victory's Fellowship of Ministries, which includes 800 ministers and more than 350 churches. Consider the attitude Peterson reflects: "I'm in the ministry of helps, for I am helping Pastor Billy Joe and Sharon accomplish their vision. It's our belief that Jesus personally 'hand picks' a senior pastor to be the leader of that ministry and congregation, so those of us working under that senior pastor must take his God-given mindset. I first must have the mind of God, I naturally have the 'mind of Jerry,' but I also must have the 'mind of Daugherty.' When I face a situation, I ask myself, 'How would Pastor act in this situation? What would he do? What would he want me to do?'"

"That gives the senior pastor another set of 'eyes' in the congregation," Peterson emphasized. "When I spot someone in the congregation that I sense has leadership or ministry giftings, I find out what his interests are, and I get him plugged into a cell — whether it be music, children, business, men, women, or whatever. I don't have 'tunnel vision' and recruit just for my own department; I see the whole blanket of groups we offer and recruit on behalf of the entire church. If you can get each person on staff doing this, in accord with their leader, then everyone works together for a single purpose: to build God's Kingdom."

The right kind of attitudes breed the right kind of activities, focused on developing leaders. Eric Castrellon states that his job is four-fold. "Our church is a great pool to find leaders. Once I find them, then I must 'feed them' with training, books, and resources; 'focus them' and get their eyes on the harvest; 'follow them' and let them lead while I observe; and 'follow-up on them' and give them as much positive feedback as I can."

Insight #5
Stress Unity and Encourage Initiative

Victory members are to "run with" Daugherty's unique vision for reaching Tulsa through groups. No fact better exemplifies this than the cell lesson printed in each Sunday bulletin. No couple better personifies this than Jerry and Lynn Popenhagen.

Unity with spiritual authority brings blessing. "When you step into leadership, when you have your first cell meeting," Jerry Popenhagen teaches, "the anointing that's on Pastor Billy Joe and Sharon will come upon you. There is a transfer of anointing, just like the anointing on Moses went onto the seventy." Care Pastor Don Hanson puts it this way, "If a leader is open to the Holy Spirit, what happens in the larger body of the church also happens in the cells, one cell at a time. Whatever is happening in Victory in that large sanctuary can happen in that cell. We are all connected to what God is doing in the larger body."

VICTORY'S TYPICAL CELL LEADER DEVELOPING OTHER LEADERS

"I actively pray and look to identify and spot potential leaders": Often

"I regularly assign potential leaders responsibilities that allow them visibility in the meeting": Sometimes

"I actively encourage potential leaders to get into training": Sometimes

To more fully develop my future leaders:

1) I discuss important aspects of the group (32%);
2) I give specific and appropriate responsibilities (30%);
3) We pray together (29%);
4) I let him/her lead group on occasion (29%).

Source: Representative sample survey (n=105) taken July 2000

This unity is reflected in other ministries based out of Victory. Victory Bible Institute (VBI) offers a foundation in God's Word in one and two-year programs on practical ministry, worship, missions, medical missions, Spirit-led business, preschool, children, and youth — and stresses cell groups. Second year VBI students lead cell groups of first year students. Faculty meet every morning in their own cell to pray for students and the day ahead.

Victory not only stresses the unity with spiritual authority, but also encourages individual initiative. Only in this way could Radhika Mittapalli start a growing young professionals' group that has generated a monthly outreach meeting in an auditorium in Tulsa's tallest downtown tower. Only because they were encouraged could Jim Husong start a group in his machinist workshop, Samantha Franklin a group in her government office, and Mark Atwell a group in his used car lot. Only with this initiative could Steve and Wendy Pogue have an idea for a young couples' ministry that has now touched hundreds of lives.

Insight #6
Be Aggressive in Reaching the Lost

Victory encourages initiative in reaching the lost more than any other activity. Victory values a person's salvation above that person's ability to mingle with the 'right crowd,' to say the right religious words at the right time, or to give large tithes and offerings. Many of Victory's cells, like bus route cells and Kidz Clubs, are aimed primarily at reaching the lost. Cell leaders of groups directed toward church members not only encourage others to bring unchurched friends and

COMMON TYPES OF CELL EVANGELISM

Empty chair evangelism provides a focal point for cell members to "release their faith," believe and pray for more to join the group.

Lifestyle evangelism starts when each member identifies one or more people they know who have challenges, and begin by praying that God will intervene and cause that person to open his or her heart to the Gospel. One is to develop a friendly relationship with each unbeliever targeted in prayer, listen with compassion, and, when appropriate, share spiritual answers to his or her problems. After a period of time, invite him or her to one's cell group, where there is even more Christian support.

Servant evangelism is serving, giving of yourself to help someone else. When acts of kindness are done, a person may want to know more of God's unconditional love, and be more receptive to Christ. This has many avenues, ranging from hot dog giveaways, to trash pickup to washing cars, to giving out free coffee at bus stops.

Monthly crusades take place in every district, and include servant evangelism. Often there is free food, free clothing, and a free medical clinic as well as direct sharing of the Gospel.

Source: *Victory's Cell Leader Manual and Training Notebook*, p.63-64.

> **OUTREACH HABITS OF VICTORY'S TYPICAL CELL LEADER**
>
> **"I actively encourage group members to invite and bring others to the group":** Regularly
>
> **When a person visits group for the first time, person is contacted:** Within a week
>
> **Number of new people leader invited to group this past month:** 3
>
> **Number of unchurched or lost people leader invited to group this past month:** 1
>
> **Group activities, events or approaches with best fruit in evangelism:**
> 1) Really, just individuals praying and inviting unbelievers to our regular gatherings (30%)
> 2) Food events (16%)
>
> Variety of other approaches with small percentages
>
> **Source:** Representative sample survey (n=105) taken July 2000

acquaintances to regular meetings, but also have creative events with food and fun to attract the hesitant. Districts hold monthly "crusades" where several groups join together, in an effort to reach the lost for Jesus. This focus helped Victory's cells report more than 6,000 salvations in 2000.

Victory's aggressive outreach is best embodied in its senior pastor. In 1998, Daugherty read in the paper that Tulsa's highest crime rate was in the 46th Street and North Cincinnati area. Victory had begun its monthly outreach crusades in Tulsa's government projects in 1989, and they had great impact. But that was no longer enough. So Daugherty gathered his wife Sharon and their children in the car and drove out to that corner area. They prayed, "God, give us land in this area for an outreach center to turn the tide of violence and to make this a place of peace."

Within a few days, two sisters contacted Daugherty. Their grandfather owned an 80-acre farm in the 46th and North Cincinnati area, and they wanted to see the land used for an outreach center to the community.

Wendell and Gloria Hope, who formerly coordinated the monthly outreach crusades and cell groups in low-income areas, now oversee Victory's 'Tulsa Dream Center,' an 18,000 foot building for food and clothing distribution, recreation for children, job training and placement, counseling, and ministry areas. Its theme? "Helping people see their dreams come true." Hope declares: "For any group, including a cell, to vitally endure, it must reach out."

This is of such importance that it is stressed in Victory's *Cell Leader Manual and Training Notebook,* which challenges leaders to ask themselves this question: "What works am I now doing that will last for eternity?" That question is followed by a poem, credited to Peter Enns, that expresses Victory's attitude:

Don't rest in the nest
Find your place in the race
Set a goal for your role
Keep your eyes on the prize

INSIGHT #7
MODIFY AND DIVERSIFY TO MEET NEEDS

Victory keeps its "eyes on the prize" of reaching the lost. Victory treasures its diligent and dedicated zip code leaders, but discovered that target groups would help reach people untouched by home groups. Victory then birthed a focus on target groups that has permeated Victory's strategy to this day — while still honoring zip code cell leaders.

When Victory discovered the "meta church" approach, and later G-12 groups, they felt both approaches would better help them build God's Kingdom and wisely adopted them. Such is the attitude that has helped birth Victory's 35 different types of groups. Now, Victory even has a limited number of special interest groups that range from clown troupes, to medical students, to those who help with clothing outreach, to those who form a soup kitchen.

What they do in training and with their groups is not unchangeable. If one schedule of cell leader training is not convenient for everyone, they add other times, including a Saturday day-long compressed training. When some leaders do not seem adequately trained, they modify and add other training.

When rigid Pharisees complained that Jesus' hungry disciples were unlawfully picking grain on the Sabbath, Jesus declared this dynamic principle: "The Sabbath was made for man, not man for the Sabbath" (Mark 2:27). Victory takes this same principle for cells: "Training, group types, and a group system were made to serve people, not the reverse. Modify and diversify training, group types, and your system to meet peoples' needs."

Insight #8
Use Wisdom in Transition

So when change is needed, what's the best approach? Victory is sensitive to leaders and others when introducing change. After Daugherty's first failed attempt at home groups, many had soured to that idea. So before the second start of home groups, Victory first took one year with monthly "Family Ministry Luncheons," when the staff would visit 100 host homes to eat, fellowship, and explain the benefits of home groups. The congregation was then positive about the 29 home groups later started, a credit to a wise approach in transition.

Remember Jeremy Baker? Before he started Kidz Clubs, he rallied 26 adults to hold three-day Backyard Bible Clubs. Only after those clubs went well did he challenge them to disciple the children they had reached through ongoing Kidz Clubs.

When Daugherty wanted to start G-12 leadership groups, he found some of his staff resistant. Instead of dictating that change, he spent weekly time with the staff for nine months, and gave them material to read and consider, highlighting the benefits of G-12 groups. When Care Pastors started their own G-12 leadership groups, they still allowed many of their area coordinators to continue in that role, and they are gradually making the transition, one person at a time.

HOW TO TRANSITION YOUR SUNDAY SCHOOL INTO SUNDAY CELLS

- **Model cell ministry** to your teachers; if you are over these teachers, invite them to your cell group.
- **Ask teachers** how they could encourage students to begin their own cell groups.
- **Help each teacher** view each member as a potential leader of another Sunday cell or other kind of cell group.
- **Encourage teachers** to allow cell group testimonies during class time.
- **Encourage teachers to visit** members' cell groups on occasion, to encourage them.
- **On a weekly basis teachers** should encourage class members to start groups.
- **Frequently teachers** should have class members pray for one another, thus empowering them to share ministry in the classroom.

Source: Based on material by Gary Stanislawski, M.C.E.
Victory Cell Ministry Resource Book, p.38.

INSIGHT #9
FOCUS ON FAMILY VALUES, AND UNDERSTAND THE BENEFITS OF HOMOGENEOUS "TARGET" GROUPS

Look at Victory's monthly directory of cell groups, and see staff pastors listed with both the husband's and wife's names. Attend Victory's dynamic worship services very long, and hear Billy Joe and Sharon Daugherty 'team preach.' Hang around pastoral staff for any period of time, and watch couples who minister together. Attend zip code cells led by couples, several with provision for children through Kidz Clubs, that meet in different rooms of the same homes.

Victory's protocol is that men minister to men, and women to women. This protocol permeates cell meetings and church services, followed even with altar respondents. Victory works hard at preserving the sanctity of marriage, and encourages family values in both the church and Victory Christian School, now with 1,500 students.

While Victory honors the mixed group that reflects family values, Victory also gives room to the homogenous "target" group, where members share a common trait like gender, age, or profession. Take one of the ladies' groups as an example: communication flow is easier in an all female group, with clearer focus on whom one should reach — another woman, something not as clear in a mixed group.

When I took a closer look, I found that Victory's target groups have grown faster and more easily than their mixed groups. I grew up both in West Africa and Korea; the third world has known for some time that homogeneous 'target' groups, especially gender-based ones, tend to be healthier and grow better. Consider Dr. Cho's church in Seoul, Korea, César Castellanos's church in Bogotá, Colombia, and Dion Robert's church in Abidjan, Ivory Coast. With the exception of 400 of the 20,000 cells in Bogotá, all the cells of these three churches are homogeneous and gender based, men meeting with men and women with women. Victory has also discovered that principle and added homogeneous target groups; its system has reaped the benefits.

INSIGHT #10
LET FERVENT PRAYER PERMEATE

But all Victory's groups must have fervent prayer to flourish. One of my favorite times at Victory was on Tuesday at 12:30 P.M., as Victory's department heads joined their senior pastor for a weekly time of fervent prayer. Daugherty started with words of encouragement, took requests, and prayed for specific cell situations. Then Associate Pastor Bruce Edwards distributed nearly 3,000 written prayer requests, taken from the flaps of the previous Sunday's offering envelopes. Each staff pastor took a stack of "request flaps;" after praying aloud for a request, he put his initials in the corner, going then to the next request, until he had prayed for each request in his stack. These were not quiet, respectable prayers, but loud and spirited; there was a tangible sense of God's presence in that room.

Fervent, faith-filled prayer must permeate every cell meeting with prayer prioritized and practiced by the staff and cell leaders. Margaret Hawthorne, director of Victory's Prayer Department, points out the best way to learn to pray: simply be with someone who prays God's Word in strong faith. She encourages each cell leader to find and

RESPONSIBILITIES OF THE CELL GROUP PRAYER LEADER

Victory encourages each zip code cell group to have a "prayer leader," under the direction of the primary leader, as part of the leadership team. The responsibilities of a prayer leader are:

- **Faithfully attend** your zip code cell group.
- **Lead the group prayer time,** when requested by the group leader, including Victory's monthly list of prayer topics when possible.
- **Keep a "prayer journal" for the group,** recording both prayer requests and praise reports of answered prayer.
- **Receive urgent prayer requests** from the group between group meetings, then phoning group members to join in prayer for those requests.
- **Conduct "prayerwalks"** in the neighborhood surrounding that group's host home.
- **Canvass that neighborhood,** offering to take prayer requests from neighbors with needs or concerns.
- **Personally intercede** for your cell group leader.

Source: Separate handout prepared by Margaret Hawthorne

designate someone in his cell to be a 'prayer leader,' and she challenges all prayer cell leaders to teach loudest by their own examples.

Some Care Pastors suggest that each new leader find at least five people to pray regularly for his group. Victory encourages leaders to keep a prayer journal during group meetings, so that they will be able to persist in prayer and record God's answers. "Prayer is primary," one staff pastor told me, "but sometimes we're just not desperate enough. Leaders must love people enough to be as desperate as they are. Every time you have a cell meeting, look for the gifts of the Spirit to operate, look for lives to be changed."

"Prayer," Daughtery often repeats, "moves the hand of God and releases His blessings. Prayer stops the power of the enemy. Prayer changes the world, changes circumstances, changes us." Standing again before a gathering of preachers, Daugherty declared, "You must have fervent prayer. If you are going to move your church off dead center, off the mediocrity it has been in, there has to be a band of people, however small, given to fervent prayer. Only fervent prayer will knock out mediocrity. In prayer you must deal with the things that have happened, the bondage, the tradition, and the heaviness that drags a church down."

PRAYER LIFE OF TYPICAL LEADER AND GROUP

Prays for each group member by name: Maybe once or twice a week

Time spent in daily personal prayer and devotions: 30-45 minutes

Time spent in prayer in most meetings: 10-15 minutes

Intentionally takes time to share answers to prayer at each group meeting: Yes

Group sees wonderful answers to prayer: Regularly

SOURCE: Representative sample survey (n=105) taken July 2000

INSIGHT #11
CONTEND FOR THE MIRACULOUS

Victory's focus on prayer and God's Word forms the basis for their focus on the miraculous. One of Victory's visitor's brochures even has the heading, 'A place were miracles happen.' Daugherty contends that when one fully believes God's Word, one will also believe in God's miracle power.

Talk to any Victory cell leader, and they will agree. Victory cell leaders pray and believe for God's healing and miraculous intervention in people's lives. I often found myself amazed and delighted at how quickly cell leaders would tell about various healings and miracles that took place during meetings or in the lives of group members. One's baby had a seizure during a cell meeting, which instantly stopped when the group prayed; another had documentation of a cancer that doctors said they could no longer find; still another told of freedom from years of addiction to drugs.

> **CELL GROUP TESTIMONY**
>
> **Richard Butler** killed a man in self-defense. When the State released Butler from prison in 1994, he found his way to Tulsa and became a student at Victory Bible Institute, learning the benefits and inner workings of Christian cell groups for the first time. Richard met and married at Victory, and soon he and his new wife were leading a cell. During the next three years, they saw God do marvelous things in their cell group: salvations, healings, deliverances, and prodigal children returning home.
>
> **Butler declares,** "God is good. God can take something out of nothing and create what He wants. As Sharon Daugherty sings, 'God has taken me out of darkness into His marvelous light.' Of all the things that have happened in our cell group, I am most grateful for the many lost people who have come to Jesus. When others see what we have, they want the same thing. That's what cell groups are all about."

Daugherty enjoys pointing to scriptural example when he talks of the importance of miracles. How did the 3,000 and 5,000 of Acts 2 and 5 come to the Lord? The 3,000 heard the disciples speaking in known languages the disciples did not understand (Acts 2:4-13). Once that miracle caught their attention, Peter preached, and only then did the people respond (Acts 2:14-41). Peter preached again in Acts 3, but not until after the lame man had been healed (Acts 3:1-10), and 5,000 believed (Acts 4:4).

"When the crippled are healed and the blind see," Daugherty highlights, "'peoples' lives are changed. Contend for the miraculous — miraculous salvations, miraculous deliverances of drugs, of alcohol and of bondages. People have had enough of the 'do's' and 'don'ts' of religion, but when they see their little crippled child walk, when they have sure conviction that God is alive, they'll easily give up the sin they have held in their lives. We must pray and press into a release of miracles, signs and wonders."

Sometimes, Daugherty explains, people learn and move in God's power in unconventional ways. I remember the moment when Daugherty looked over a seminar crowd of ministers before him, paused and declared with resolve of soul, "It's easier to control wildfire than it is to raise the dead. We must let God have His way in our churches and in our cell groups."

Insight #12
Tailor Your Cell System to Your Church's Culture

Pastors say many things about cell groups, especially why they do or do not work in their churches. The negative comment I hear most often is, "Cells are cultural. They'll work well in that country because of its culture, but not in America."

Problem is, that statement's not true. Victory has clearly shown that cell groups are based on biblical principles that transcend culture. No matter what our culture, we human creatures share similarities. We want to be loved and valued. We want to achieve. We long to belong.

Cells meet some of our deepest needs. Cell groups form prime vehicles to express the priesthood of the believer. Being a cell group leader is the best and most realistic experience many will have in ministry.

So why do some churches flourish in cells while others flounder? If it is because of the culture of one's country, then why are there some churches in Korea that never got cells to work well, while others in America have? Perhaps it has more to do with the "culture" of that church than the culture of that society.

Consider Victory's "culture." As a church, Victory prays hard and understands and practices spiritual authority. The staff and leaders align themselves with the house vision and sacrifice personal preferences for the greater cause. Victory's culture stresses the believer's ministry and makes heroes of everyday Christians who pray, believe God, and lay hands on the sick. Victory's culture puts priority on aggressive outreach and leaves little room for the uninvolved. I even heard about one woman who left Victory to find a church where she could "just sit and be comfortable." Victory had too much challenge for her.

Spiritual authority is easier to follow when your senior pastor has a proven track record of godly choices. It's easier to believe that God can use you to minister when your pastor and staff continually make heroes of believers who have seen God heal and work through them. You too might be more bold in evangelism if you were an active leader of a group focused more on the lost, maybe a business group in a secular office.

Daugherty is a man of integrity, driven by godly values and beliefs. When he valued what he saw at Dr. Cho's church in Korea in 1983, he returned home with renewed enthusiasm for home groups and the believer's ministry. When God dealt with him about reaching the receptive and poor in 1987, he threw all of his energies in that direction, even if it meant the downslide of the home groups. It took years before Daugherty could see how cells fit with his aggressive outreach to the receptive and poor.

When Daugherty refocused on groups and decided that each department would have its own cells, things changed. At that point Victory adopted what I term the "inclusive approach." Each leader of any group or grouping was "included" as a cell leader if he would do at least three parts of the five-fold vision and go through required screening and training. These groups emphasized increased development of relationships and giving ongoing care to group members, while keeping focus on reaching the lost.

Did Victory compromise with the inclusive approach? Maybe to cell church purists who demand that cells only meet in homes and that all significant lay ministry be done only through those home groups, it did. If you feel the same way, spend a week with me at Dr. Cho's church. As wonderful and powerful as their home groups are, you would discover that not all their significant lay ministry happens just through their home cell groups. Sometimes cherished ideals blur reality.

Daugherty and his staff have done the only thing their integrity would allow. Daugherty was not going to give up aggressive outreach to the receptive and poor. He wanted Sunday School classes to continue while he still believed in the importance of home groups. So all were included in the "house vision" for cells.

As a result, Victory developed a more unified training and tracking system.

Now Victory's varied leaders feel themselves on equal ground. I remember Lynn Popenhagen telling about the head nursery coordinator, who didn't know how she and her workers fit in Victory's cell vision. Then she realized that her nursery groups did three out of five parts of the vision. They also developed relationships with children and their families and saw many come to the Lord. That nursery coordinator's despair turned into delight. She found that she and her leaders fit into Victory's vision for cells.

Your situation might be different. You might need to give a total focus to home cell groups. Maybe you're planting a church in a suburban area with a congregation core that hungers for a sense of stable Christian community through home groups. Or you might have an inner city church where fragmented lives can only be cemented in home cells.

Know your church culture, my friend. Most of all, obey your God. Allow His Spirit to lead you on an adventure, using creative cell groups as vehicles to break through and reach every segment in your city!

CHAPTER EIGHT: SMALL GROUP DISCUSSION

Icebreaker: Describe an insight you learned about life while you were growing up. How has that helped you as an adult?

Questions about this chapter:

1. In your own experience, which one or two of the insights in this chapter have you found to be true?
2. According to these insights, what should be the senior pastor's view of his responsibilities in a cell system? What about staff pastors? What about a leader in his or her own cell group?
3. Was there any insight that surprised you? Why? Which one or two insights did you consider most important? What are your reasons for that choice(s)?
4. If you were to choose the three most helpful insights from this chapter, which ones would you select? Why?
5. By this point you have probably read most of this book on Victory. If you were to add one or two of your own distinctive insights to this list, what would they be?

Application: Share one principle or practice you gleaned from this chapter on insights into Victory's cell system. How could that be best integrated into your present or future cell group? Your church?

Concluding Challenge:

Understand Our Times

"And of the children of Issachar, which were men that had understanding of the times, to know what Israel ought to do; the heads of them were two hundred; and all their brethren were at their commandment" (1 Chr. 12:32).

Sharon Daugherty lived in the rice country of southwestern Arkansas, and later in the cotton and soybean farmland along the Arkansas' delta region. Billy Joe and his wife preached often in that area during earlier days of their marriage and talked to those farmers about harvest time.

Harvest time for cotton or soybeans only lasts about two weeks. When that crop ripens, those farmers work hard. They work from before sunrise until it gets dark, then turn tractor lights on and reap the harvest until midnight. After four to five hours sleep, they get up and start the same process again.

In the days when Billy Joe and Sharon ministered in Arkansas, the economy of small rural communities depended on those crops, so during harvest time they shut down the grocery store, the bank, and all of the major businesses. Why? Because if the harvest was not reaped, they weren't going to need a grocery store, bank, or business. There's only a short window of opportunity to reap a harvest when it is ripe; if you do not reap then, you will lose it. So everyone focused on the harvest.

"Most people in America don't have a clear concept of end-time harvest," Daugherty explains. "It's not just Billy Graham or full-time evangelists that are needed — ALL of the workers are needed in the fields. We don't want any part of the harvest lost. We need every person possible working in the fields to bring the harvest in.

"We are in a crucial window of opportunity," Daugherty emphasizes. "Remember Jesus' words, 'Lift up your eyes, and look on the fields, for they are white already to harvest' (Jn. 4:35). If the harvest was ripe 2000 years ago, what is it now? Harvest is here right now, and end-time harvest will intensify. Half of the six billion people in our world have never even heard the name of Jesus," Daugherty points out. "Because of the exponential factor in the population explosion in the twenty-first century, almost as many people can die and go to hell in our generation as have died and gone to hell in all of past history."

Noah did not start building the ark after the rains began to fall. Instead, 100 years before, God told Noah what was coming (Gen. 5:32-7:11). Noah had to build that ark without seeing any rain. The same flood that lifted and saved Noah and his family because they had prepared, destroyed the wicked because they were not ready. The coming flood of end-time harvest is going to have two different effects, depending on what people have done to prepare.

Like the men of Issachar, we must understand our times. We're in an hour like no other. Consider your community. What percentage don't yet go to church? How many are not yet born again? How could a diversity of creative cell groups best break through and reach them?

"We must build the cell ministry to prepare for the coming flood of end-time harvest," Daugherty contends. "Many pray for revival, but few prepare for it."

Each church must have God-birthed and Spirit-breathed goals, vision, and revelation to drive more workers into the harvest. Your preparation is evidence that you believe your prayers have been heard and answered.

"Like Elijah in 1 Kings 18," Daugherty adds with fervor, "I can hear the sound of abundance of rain. Like Elijah told his servant, go look, go look, go look again and again. Like the promise of downpour in a cloud the size of a man's hand, it is coming, it's coming upon the earth."

What does God want your church to do about cells? What does God want you to do about the flood of end-time harvest? It is coming, it's coming upon the earth. What would God have you do?

INDEX

BOOKS TO ADVANCE YOUR CELL MINISTRY

HOW TO LEAD A GREAT CELL GROUP MEETING . . .
. . . So People Want to Come Back
by Joel Comiskey
Joel Comiskey takes you beyond theory and into the "practical tips of the trade" that will make your cell group gathering vibrant! This hands-on guide covers all you need to know, from basic how-to's of getting the conversation started to practical strategies for dynamic ministry times. If you're looking to find out what really makes a cell group meeting great . . . this book has the answers! 144 pgs.

LEADERSHIP EXPLOSION
Multiplying Cell Group Leaders to Reap the Harvest
by Joel Comiskey
Cell Groups are leader breeders. Yet few churches have enough cell leaders ready to start new groups. In this book, you will discover the leadership development models used by churches that consistently multiply leaders. Then you will learn how to create your own model that will breed leaders in your church. 208 pgs.

HOME CELL GROUP EXPLOSION
How Your Small Group Can Grow and Multiply
by Joel Comiskey
This is the most researched and practical book ever written on cell-group ministry! Joel traveled the globe to find out why certain churches and small groups are successful in reaching the lost. He freely shares the answer within this volume. If you are a pastor or small group leader, you should devour this book! It will encourage you and give you simple, practical steps for dynamic small group life and growth. 152 pgs.

GROUPS OF 12
A New Way to Mobilize Leaders and Multiply Groups in Your Church
by Joel Comiskey
Finally, the definitive work that clears the confusion about the Groups of 12 model. Thousands of pastors have traveled to International Charismatic Mission to see it in operation. In this new title, Joel has dug deeply into ICM and other G-12 churches to learn the simple G-12 principles that can be transferred to your church. This book will contrast this new model from the classic structure and show you exactly what to do with this new model of cell ministry. 182 pgs.

Order Toll-Free from TOUCH Outreach Ministries
1-800-735-5865 • Order Online: www.touchusa.org

SUBSCRIBE TODAY TO CellGroup JOURNAL

Equip Your Cell Leaders!

CellGroup JOURNAL

CellGroup Journal, unlike any other periodical, is focused on the needs and desires of cell leaders in your church. Every quarterly issue contains practical feature articles and columns from some of the most respected leaders in the U.S including Ralph W. Neighbour, Jr., Billy Hornsby on leadership, Karen Hurston on evangelism, Gerrit Gustafson on worship, Sam Scaggs on missions, and Larry Kreider with a closing note on a variety of topics. Pastor's get fed too . . . each issue contains an article for pastors by a pastor who has learned a good lesson in cell life and wants to share. Bulk discounts are available for larger subscriptions. Call today to subscribe for all your cell leaders and staff!

Order Toll-Free from TOUCH Outreach Ministries
1-800-735-5865 • Order Online: www.touchusa.org

CONFERENCES TO ADVANCE YOUR CELL MINISTRY

How to be a GREAT Cell Leader

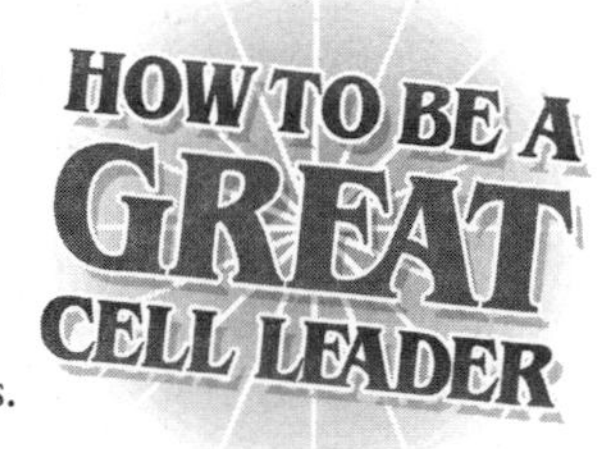

This is basic training for cell leaders! You will gain practical help in facilitating the life of your cell group both in your weekly meetings and the days in between. You will gain an understanding of the natural life-cycle of a cell and how to work within that cycle to accomplish kingdom purposes. In addition, you will receive:

- ✔ A relational, cell-based approach to evangelism.
- ✔ A mentoring-based strategy for growing disciples.
- ✔ Plenty of time to ask questions and dialogue with other cell leaders and interns.
- ✔ Small group interaction and role play.

Sessions Include: From Dream to Reality: An Overview of the Cell Church Movement • What is a Cell Group? • Profile of a Cell Leader • The Life-Cycle of a Cell • Friendship Evangelism: Reading Your World Through Cell-based Evangelism.

The afternoon session, "DynamicCellMeeting.com" will be filled with exciting interactive learning, covering such topics as: How to Facilitate a Cell Meeting • Creating a Transparent Environment • How to Ask Good Questions • Listening to Understand • Youth and Children in the Cell • Leading Worship in the Cell • Ministering to Difficult People • And much more!

Exploring the Cell-Based Church

You have heard about cell ministry and wondered about its effectiveness. Will it work well in your setting? This one-day seminar is designed to give you a comprehensive introduction to cell-based ministry on an introductory level. After the seminar, you will be equipped to make an informed decision as to whether cell-based ministry will work in your unique setting. You'll also know if you should investigate the model further.

Who should attend? Pastors, staff members, and other interested church leaders.

Sessions Include: The Way It Is • The Way It Can Be • The Answer • How It Works • U.S. Cell Church Models • Where Do You Go From Here?

Register Today! 1-800-735-5865
Register Online: www.touchusa.org

268.64
H9669
100582

CONFERENCES TO ADVANCE YOUR CELL MINISTRY

LINCOLN CHRISTIAN COLLEGE AND SEMINARY

Designing the Cell-Based Church

Can cell-based ministry work in your setting?

Absolutely! This conference will provide you the needed foundation for building a successful cell ministry in your own church setting.

You will learn cell-church principles and values, cell-church structures and systems, and theological and theoretical foundations for cellular ministry.

You will gain a historical perspective of the cell church movement, and be introduced to U.S. churches of all sizes who are effectively using the cell model to reach and disciple followers of Jesus.

You will learn effective cell-based methods of evangelism, discipleship, and body-life development.

And, you will be given expert help in assessing your church's readiness for adopting cell-based ministry and designing a cell-based ministry model for your church setting.

Overcoming Barriers to Cell Church Growth

Every church faces growth barriers. This conference will help you address those, especially those unique to the cell church environment.

Learn how to:

- ✔ Jump-start stagnated cells
- ✔ Multiply cells
- ✔ Use a church planting approach to cell multiplication
- ✔ Facilitate youth cells for Kingdom growth
- ✔ Enhance body-life in your cells
- ✔ Coordinate corporate and cellular evangelism efforts
- ✔ Deal with children in the cell
- ✔ Offer person-centered discipleship that produces disciples who produce disciples

Register Today! 1-800-735-5865
Register Online: www.touchusa.org

3 4711 00154 7613